ETHICS IN ENGINEERING

ETHICS IN ENGINEERING

Mike W. Martin
Department of Philosophy
Chapman College

Roland Schinzinger
School of Engineering
University of California, Irvine

McGraw-Hill Book Company
New York   St. Louis   San Francisco   Auckland   Bogotá
Hamburg   Johannesburg   London   Madrid   Mexico   Montreal   New Delhi
Panama   Paris   São Paulo   Singapore   Sydney   Tokyo   Toronto

This book was set in Optima by Black Dot, Inc. (ECU).
The editors were Kiran Verma and David A. Damstra;
the copy editor was Lester Strong;
the production supervisor was Charles Hess.
The drawings were done by ECL Art Associates, Inc.
The cover was designed by Jo Jones.
R. R. Donnelley & Sons Company was printer and binder.

Cover photograph: © Barrie Rokeach 1983.

ETHICS IN ENGINEERING

2 3 4 5 6 7 8 9 0 D O C D O C 8 9 8 7 6 5 4 3

ISBN 0-07-040701-0

Library of Congress Cataloging in Publication Data

Martin, Mike W.
     Ethics in engineering.

     Bibliography: p.
     Includes index.
     1. Engineering ethics.   I. Schinzinger, Roland.
II. Title.
TA157.M33 1983   174'.962   82-18697
ISBN 0-07-040701-0

For Shannon, and for my parents,
Theodore and Ruth Martin.

Mike W. Martin

For Jane, and for my father
Robert Schinzinger.

Roland Schinzinger

CONTENTS

PREFACE

Technology has a pervasive and profound effect on the contemporary world, and engineering plays a central role in all aspects of the development of technology. Because of this it is vital that there be an understanding of the ethical implications of engineers' work. Engineers must be aware of their social responsibilities and equip themselves to reflect critically on the moral dilemmas they will confront. The public must acquire an understanding of the extent and limits of the responsibilities of engineers; it must be prepared to shoulder its own responsibilities where those of the engineers end.

This book is written for students preparing to function within the engineering profession and for their colleagues in philosophy and the humanities who struggle to apply ethical theory to the pressing problems of everyday life within a technological society. It is also intended to increase interest in engineering ethics among practicing engineers, social scientists, philosophers, and all those engaged in the enterprises of technology, including the general public.

PURPOSE

Ethics in Engineering provides an introduction to the basic issues in engineering ethics, with emphasis given to the moral problems engineers face within the corporate setting. It places those issues within a wider philosophical framework than has been customary in the past, and it seeks to exhibit both their social importance and their intellectual challenge. The primary goal is to stimulate critical reflection on the moral issues surrounding engineering practice and to provide the conceptual tools necessary for pursuing those issues. The book is intended to be a teaching instrument while also serving to advance the field of engineering ethics.

In large measure we proceed by clarifying key concepts, sketching alternative views, and providing relevant case study material. Yet in places we argue for particular positions which in a subject like ethics can only be controversial. We do so because it better serves our goal of encouraging critical judgment than would a mere digest of others' views. Accordingly, our aim is not to force conviction, but to provoke either reasoned acceptance or reasoned rejection of what we say. We are confident that such reasoning is possible in ethics, and that through lively and tolerant dialogue progress can be made in dealing with what at first might seem irresolvable difficulties.

COURSES

Sufficient material is provided for full courses devoted to the topic. The book can also be used for a several-week module on engineering ethics within a variety of other courses in engineering, business, philosophy, and the social sciences—courses which typically include such topics as general professional ethics, business ethics, applied philosophical ethics, engineering law, business management, values and technology, engineering design, technology assessment, and safety.

OUTLINE

Ethics in Engineering is divided into four main parts which emphasize, respectively, applied ethical theory, safety and risk, the corporate setting, and career choices and further problems.

Part I provides an introduction to basic concepts and theories in ethics. Chapter 1 introduces and defines the field of engineering ethics as it relates to applied philosophical ethics. The roles of describing, evaluating, and clarifying in ethics are explained, and the goal of fostering moral autonomy in studying ethics is emphasized. Chapter 2 introduces further key concepts: moral dilemmas, responsibility, and relativism. Three fundamental types of ethical theories, emphasizing duties, rights, and good consequences, are introduced.

Part II deals with the moral issues surrounding safety assessments within the inherently risky activity of engineering. Chapter 3 develops a perspective on engineering as an experiment on a societal scale involving human subjects. This model provides a framework for discussing various aspects of responsible engineering practice: imaginative foreseeing of possible effects; careful monitoring of projects; and respecting the rights of clients and the public to make informed decisions about the products which affect them. The model also provides a context for discussing the importance and limitation of laws and codes of ethics in engineering. Chapter 4 explores some of the moral complexities of making safety and risk decisions.

Part III examines some of the special moral issues arising for the nearly 90 percent of engineers who are employed within corporations. Chapter 5 approaches those issues by clarifying the ideas of professionalism, loyalty to employers, and employer authority. It also contains discussions of employee obligations to employers related to conflicts of interest, confidentiality, and unionism. Chapter 6 focuses on the various rights of engineers: professional, employee, and human rights. Special topics include whistle blowing, freedom of conscience, due process from employers, and discrimination or preferential treatment.

Part IV sketches some further issues. Chapter 7 explores the role of morality in making career choices. Special moral issues concerning computers, military work, the environment, and consulting engineering are discussed. Chapter 8 provides a brief review of some major themes and outlines a few additional issues concerning the responsibilities of the engineering profession.

The Appendix contains sample codes of ethics of several major engineering professional societies.

ACKNOWLEDGMENTS

This book was written after but was inspired by our participation as a philosopher-engineer team in The National Project on Philosophy and Engineering Ethics from 1978 to 1980. We gratefully acknowledge the sponsor of that project, the National Endowment for the Humanities, and the project director, Robert J. Baum.

Mike Martin's work on the book was largely made possible by a fellowship from the National Endowment for the Humanities from 1981 to 1982. He is also indebted to Nelson Pike and the Department of Philosophy at the University of California, Irvine, for granting him the status of Visiting Scholar during that time.

We greatly benefited from the criticism and suggestions of participants in many of the classes and seminars we have conducted during the past several years. The formal course work ranged from segments of design courses given at the U.C.I. School of Engineering to full courses on engineering ethics taught in the U.C.I. Department of Philosophy and for the Masters-Degree-in-Engineering Program of California State Polytechnic University, Pomona, taught at Fluor Corporation. An additional seminar brought together practicing engineers under the auspices of the Institute of Electrical and Electronics Engineers, Orange County Section.

Many colleagues and friends have influenced our thinking about the issues in engineering ethics. Our special thanks extend to Frank Alderman, Jim Andersen, Robert Anderson, Robert J. Baum, Taft Broome, Robert Bruder, Richard T. De George, Paul Durbin, John Fielder, Albert Flores, Leslie Francis, Gilbert Geis, Jerry Gravander, Milton Gross, Jack Hagman, Michael Hodges, Lawrence Hollander, Elizabeth Jacks, David James,

Andrew Jameton, Edwin Jones, Kenneth Kipnis, Rob Kling, Robert Kunde, Robert Ladenson, Bruce Landesman, Edwin Layton, John Leckie, Barry Lichter, William Litle, Thomas Long, Ruth Macklin, Bill McIlvaine, Frank Manley, A. I. Melden, Robert Miles, John Mingle, Martha Montgomery, Carl Nelson, Richard Nesbit, Andrew Oldenquist, Sterling Olmsted, James Otten, David Palmer, Victor Paschkis, Susan Peterson, Michael Pritchard, Charles Reagan, Robert Redmon, Tom Riley, Tom Rogers, Jean Runzo, Joseph Runzo, Robert Saunders, Fay Sawyier, Ken Schneider, Robert Schultz, Sharon Schwarze, William Schwarze, Edward Slowter, Carol Ann Smith, Robert Smith, Sheri Smith, James Thorpe, Paul Torda, Stephen Unger, Vivian Weil, Carolyn Whitbeck, Donald Wilson, and Janet Zimmerman.

Finally, we wish to thank the many authors and publishers who granted us permission to use the copyrighted material as acknowledged in the bibliographical entries. We also thank the Institute of Electrical and Electronics Engineers, the National Society of Professional Engineers, and the Accreditation Board for Engineering and Technology (formerly the Engineers' Council for Professional Development) for allowing us to print their codes of ethics in the Appendix.

Mike W. Martin
Roland Schinzinger

ETHICS IN
ENGINEERING

PART I

THE
SCOPE OF
ENGINEERING
ETHICS

Whether or not it draws on new scientific research, technology is a branch of moral philosophy, not of science. It aims at prudent goods for the commonweal and to provide efficient means for these goods. . . . As a moral philosopher, a technician should be able to criticize the programs given him [or her] to implement.

Paul Goodman

Philosophy, though unable to tell us with certainty what is the true answer to the doubts which it raises, is able to suggest many possibilities which enlarge our thoughts and free them from the tyranny of custom.

Bertrand Russell

CHAPTER 1

INTRODUCTION

Engineering ethics is the study of the moral problems confronted by individuals and organizations involved in engineering. While the emphasis of this book will be upon engineers, the ethics of engineering is wider in scope than the ethics of engineers. It applies also to the decisions made by anyone else having to do with technological enterprise, including scientists, managers, production workers and their supervisors, technicians, technical writers, government officials, lawyers, and the general public.

The moral problems surrounding engineering should interest each of us, for we are all affected daily by engineering projects having moral dimensions. The increasing attention now being given to engineering ethics has been the result, perhaps unfortunately, of some negative effects of technology. Public awareness has been heightened by well-publicized cases such as Three Mile Island, Love Canal, DC-10 crashes, unsafe Pinto gas tanks, and farmworkers displaced by machines. Attention has also been drawn to many less obvious but nevertheless harmful effects of technology that do not impinge directly on our senses, such as those involving asbestos products and radiation.

Why has the general significance of engineering ethics, with its focus on personal decision making and responsibility, only recently become appreciated? In spite of the dramatic impact of engineering on our safety and well-being, we have tended to stereotype it as a tool of vast impersonal organizations. Individuals involved in it have frequently been viewed as cogs in machines rather than as responsible decision makers. Emphasis has usually been upon products and their effects on society rather than upon the human drama behind their production. Yet engineering products derive from personal creative activity in which responsible conduct can make the difference between large-scale benefit or large-scale harm, up to and including life and death. Engineering ethics is the discipline which examines the moral import of that creative activity. It explores the moral dimensions of technology "from the inside."

The study of engineering ethics has traditionally been too narrowly conceived. Often it has been regarded as encompassing little more than the formulation and transmission of official prescriptions by professional societies in the form of codes, guidelines, and opinions. These activities are vitally important. But they should be viewed as only one aspect of engineering ethics, not its full substance.

There are two contrasting approaches to engineering ethics, one of which emphasizes small everyday problems and the other larger social problems. Those of us who tend to view the world from the microcosm of our immediate surroundings may become preoccupied with the frequently petty, but nevertheless persistent and nagging, moral problems of everyday life and work. What we must do is to reach beyond those problems to seek an understanding of their root causes.

Others among us are more inclined toward a macroscopic view involving reflections on the moral condition of society. What we need is the discipline to reconcile the broad view with specific circumstances as they present themselves in different everyday settings.

Which approach is better? Neither by itself. What is required is ongoing interactions between the two. In this vein we will approach our subject, taking as examples some cases involving a wide social impact and others of a narrower or more routine nature.

VARIETY OF MORAL PROBLEMS

An engineered product or project goes through various stages of design, manufacture or construction, testing, sales, and service. Engineers carry out or supervise the appropriate activities at whatever stage of this process a convenient division of labor has assigned them. The nature of the activity or project will generally dictate whether the engineers involved are by training civil, electrical, mechanical, or chemical engineers, to name only the major fields, but every field involves moral problems.

For example, as engineers carry out their tasks, there will be times when their activities will ultimately lead to a product which is less than useful or safe. This may happen intentionally, or under pressure, or in ignorance. A product may be intentionally designed for early obsolescence; an inferior material may be substituted under pressure of time or lack of money; or a product's eventual harmful effects may not be foreseen. Then too, because of the size of a project, or because of the large numbers of a product sold on the mass market, many people may be affected. And these problems arise quite apart from the temptations of bribes and other forms of outright corruption.

The following four specific examples barely hint at a few of the areas covered by engineering ethics:

1 An inspector discovers faulty construction equipment and applies a violation tag preventing its continued use. The inspector's superior, viewing as minor the infraction of a relatively insignificant regulation, orders the tag removed so the project will not be delayed. The inspector objects and is threatened with disciplinary action.

2 An electric utility company applies for a permit to operate a nuclear power plant. The licensing agency is interested in knowing what emergency measures have been established for human safety in case of reactor malfunction. The utility engineers describe the alarm system and arrangements with local hospitals for treatment. They do not emphasize that these measures apply to plant personnel only and that they have no plans for the surrounding population. "That is someone else's responsibility," they claim upon being questioned about this omission.

3 A chemical plant dumps wastes in a landfill. Hazardous substances find their way into the underground water table. The plant's engineers are aware of the situation but do not change the disposal method because their competitors do it the same cheap way and no law explicitly forbids the practice.

4 Electronics company ABC gears up for production of its own version of a popular new item. The product is not yet ready for sale, but even so, pictures and impressive specifications appear in advertisements. Prospective customers are led to believe that it is available off the shelf and are drawn away from competing lines.

These examples show how ethical problems arise most often when there are differences of judgment or expectations as to what constitutes a proper course of action. The engineer may be faced with contrary opinions from within the firm, from the client, from other firms within the industry, or from government. And not to be left out, as indicated in Fig. 1-1, are still other possible positions that can be taken by the profession of engineering itself, usually embodied by or represented in the form of a professional society. We will have more to say about the relationships between engineers and their colleagues, their managers, and their clients later, particularly in connection with professional freedoms, rights, and obligations.

The four cases have already raised a number of pertinent moral questions. To what extent should an employer's or supervisor's directives be the authoritative guide to an engineer's conduct? What does one do when there are differences of judgment? Is it fair to be expected to put one's job on the line? Should one always follow the law to the letter? Is an engineer to do no more than what the specifications say, even if there are problems more serious than those initially anticipated? How far does an engineer's responsibility extend into the realm of anticipating and influencing the social impact of the projects he or she participates in?

The case of the inspector told to ignore rules has numerous variations: in the testing of prototypes or finished products in a factory, in the handling of financial audits, or in the response to unsolicited reports of dangerous situations, to name but a few.

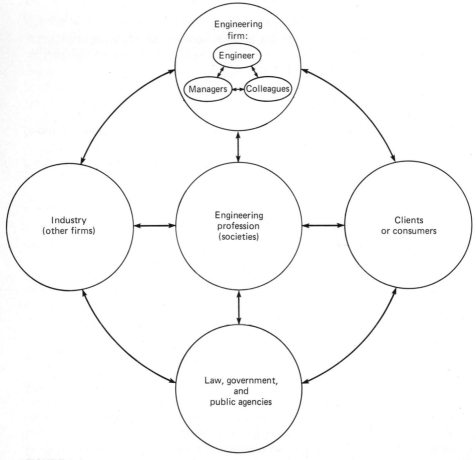

FIGURE 1-1
Contexts of potential professional disagreements engineers may encounter.

The nuclear power plant situation demonstrates several ways in which a product's ultimate use and possible failures may be overlooked by its designers. Lack of coordination in planning for its effects on everyone connected with it can lead to catastrophic results; limiting one's attention only to the narrow specifications can result in the delivery of a product that satisfies the contract but does not satisfy the needs of the customer or the public, and so on.

The questionable conduct of the chemical plant and, to a greater degree, the ABC company falls into the category of "ethics cases" which traditionally have occupied the ethics review boards of professional societies. Prescriptive codes of conduct can be established for such infractions of common decency, and for that reason those instances constitute the majority of cases in which action has been taken. This has encouraged the identification of engineering ethics with a finite set of specific maxims and regulations designed to assure moral conduct. Yet as we shall see, the moral problems arising in this field are not always manageable in this straightforward manner.

Moral Problems and Moral Reasons

We all have some intuitive notions regarding what constitutes a moral problem. It is not easy, however, to characterize in a comprehensive way what makes something a moral problem. A good beginning is to define a moral problem as one which is resolved by a moral judgment. But what is a moral judgment?

A quick glance at the preceding illustrative cases might suggest that a moral judgment is a statement about what *ought* or *ought not* to be done in a given situation, what is *right* or *wrong* about the handling of it, or what is *good* or *bad* about the actions of the people involved in it. But saying this is not enough. There are many nonmoral uses of these words and ideas. In order to start a car a person *ought* to put the key in the ignition—that is the *right* thing to do. People *ought* to brush their teeth before leaving for school or work, and it is *good* to avoid drinking so much coffee that one becomes jittery. To increase profits a company *ought* to try to cut unnecessary costs while increasing efficiency. None of these judgments would typically be counted as moral ones. Moral judgments are about what *morally* ought or ought not to be done, what is *morally* right or wrong, and what is *morally* good or bad. Since the word "morally" is but the adverbial form of the adjective we are trying to define, we are caught in a circle if we cannot move beyond these general evaluative notions.

Some progress can be made if we characterize moral judgments in terms of the particular sorts of reasons used to justify them (Frankena, 1973, 110). This is because such reasons will typically differ significantly from the grounds we give in justifying other types of value judgments. If we ask why persons ought to brush their teeth, the answer will be in terms of health and social etiquette. If we ask why a certain painting is judged to be a good one, the answer will be in terms of its striking lines, color, unity, symbolism, and so on. Giving these reasons in support of judgments makes it explicit that the judgments are nonmoral ones. Thus, too, if an engineering design is said to be a good one simply because it is simple, elegant, or cost-effective, we know that technical and business values are at issue rather than specifically moral reasons.

What, then, is a *moral reason?* Answering this question by providing a unified account would be a complex task. Clarifying moral reasons requires clarifying what it is to act, think, and feel in morally reasonable ways. And the search for an adequate set of defining criteria for such moral reasonableness has been the central goal of philosophical reflection on morality since Socrates (Melden, 1967, 5).

In Chap. 2, several of the more influential philosophical ethical theories will be outlined. For the present, a mere list of examples of moral reasons may suffice. These can be organized into three categories: obligations and duties, rights, and ideals.

First, moral reasons are often formulated in terms of moral obligations and duties. For example, it is wrong to poison a community's water supply *because* we all have a general obligation not to harm others. Other general obligations we have, both in professional and private life, are those to keep promises and contracts; to show gratitude for personal favors we receive; to avoid causing unnecessary offense and pain; to avoid lying, stealing, and cheating; and to be fair and just. In addition, as we shall see, engineers in the course of their work acquire special duties in regard to safeguarding the public, keeping proprietary information confidential, and so on.

In ordinary speech, we should also note, the terms "obligation" and "duty" are often used interchangeably. Sometimes, however, "duty" refers to the special moral responsibilities involved in a job or profession.

Second, some moral reasons refer to moral rights as opposed to obligations and duties. Roughly speaking, rights are entitlements, either to be allowed to pursue one's private interests without interference from others or to be provided with various positive benefits. General human rights are those which people have simply as individuals, such as the rights to life, liberty, and the pursuit of happiness. There are also more specific moral rights which may be acquired in various ways. For example, when people form contracts they secure rights to have the other party keep its end of the bargain. Or again, citizens may acquire special moral rights to benefits provided by their society or government.

Third, some moral reasons refer to moral ideals. These are standards that go beyond the requirements of fulfilling basic duties and recognizing other people's rights. For example, one ideal is to support and show a high degree of concern in helping others satisfy their needs, while another is to be fully compassionate and sensitive to the suffering of others. There are many ideals involving selfless devotion to public causes like ending world hunger, promoting world peace, and helping to advance disadvantaged groups.

Which ideals one will pursue is a highly personal dimension of morality. Yet some ideals are especially important within professions, and one of our aims will be to see why the engineering profession should be concerned to foster certain moral ideals among its practitioners.

Study Question

1 Cite examples of ethical problems which can arise in the production of a new type of motor bicycle as it passes through the stages of design, manufacture, testing, sales, and service. Use your imagination in developing the details of at least one problem at each stage.

NORMATIVE, CONCEPTUAL, AND DESCRIPTIVE INQUIRIES

The study of the moral problems in engineering can be viewed as involving three distinct kinds of inquiry: *normative, conceptual,* and *descriptive.* Characterizing each will provide a fuller specification of the goals and content of engineering ethics.

First we shall discuss the notion of normative inquiries.

The primary aim of engineering ethics is to identify and justify the moral obligations, rights, and ideals of individuals and organizations engaged in engineering. It seeks to establish which prescriptions of basic duty and higher ideal ought to be endorsed on moral grounds, and attempts to apply those results to specific situations in a manner that yields practical guidance.

For example, it asks the following: How far does the obligation of engineers to protect public safety extend in given situations? What are the bases of engineers' obligations to their employers, their clients, and the general public? When, if ever, should engineers be expected to blow the whistle on dangerous practices of the employers for whom they work? Is whistle-blowing a minimal moral duty or more a matter of heroism that goes beyond the requirements of an engineer's basic duty? Whose values ought to be primary in making judgments about acceptable risks in a design for a public transport system: those of management, senior engineers, government, voters, or some combination of these? Which particular laws and organizational procedures affecting engineering practice are morally warranted? What moral rights should engineers be recognized as having in order to help them fulfill their professional obligations?

Because answering such questions calls for the giving of sound moral reasons concerning the norms or standards which ought to govern engineering practice, they are referred to as *normative* or *evaluative questions,* and the study of them constitutes a *normative inquiry.*

Dealing effectively with normative issues, however, requires two further types of inquiry. One of these is *conceptual,* directed toward clarifying the basic concepts or ideas, principles, problems, and types of arguments used in discussing moral issues in engineering. For example, a conceptual inquiry might seek to enrich and sharpen our understanding of such diverse and possibly conflicting entries in codes of ethics as the following: Engineers should "protect the safety, health, and welfare of the public," while they should also "act as faithful agents or trustees for their employers or clients."

The third type of inquiry seeks to uncover factual information bearing upon conceptual and normative issues. Since it is concerned with describing what the facts are, it is called a *descriptive inquiry.* Where possible, researchers attempt to conduct descriptive inquiries using proven scientific techniques. Topics of special interest in the context of our discussion are the business realities of contemporary engineering practice, the history of the engineering profession, the effectiveness of professional societies in fostering moral conduct, the procedures of making risk assessments, and psychological profiles of engineers. Determining the facts in these areas provides an understanding of the background conditions which generate moral problems. It also enables us to deal realistically with alternative ways of resolving those problems.

Interrelatedness of the Three Inquiries

To summarize, engineering ethics involves three distinct types of inquiry:

1 Normative, whose practical aim is to provide reasoned evaluations of the conduct of individuals, the functioning of organizations, and the alternative responses available to solve concrete problems. Interwoven with this is the more theoretical aim of justifying the major moral principles which ought to be affirmed by individuals and organizations involved in engineering.

2 Conceptual, concerned with clarifying basic ideas, principles, issues, and types of argument concerning the moral problems in engineering.

3 Descriptive, which seeks to provide factual information needed for understanding and dealing with both conceptual and normative issues.

These three types of study are complementary and usually closely interrelated, as the following example illustrates. Consider a young engineer who becomes convinced that the level of pollutants her company is pouring into a stream is dangerously high, given that children are using the river downstream for swimming. She expresses her view to her immediate supervisor, who says her fears are unfounded because the pollution has caused no complaints in the past. Is she required to do more?

Obviously this question poses a practical normative issue. But we will also want to know additional facts about the case, such as the nature of the pollutants in this instance. Then we should want to know about the costs of controlling them, and whether the engineer's conviction that they are dangerous is well founded. Hence a descriptive inquiry is called for.

Perhaps enough has already been said, however, to suggest that the engineer definitely should take further action, since her professional judgment has been overruled without a compelling reason. But should she merely request of the supervisor a more detailed response? Should she go above the supervisor to higher management? Should she write to the local mayor, or would going outside the normal organizational channels constitute disloyalty to the company? And what is the basis for holding that she even has the right to a cogently reasoned response from her supervisor?

These are further normative questions since they call for making moral evaluations of her responsibilities and rights. Answering them will require an understanding of the general moral obligations of engineers. And if this understanding is to be more than an undefended opinion or intuition, it will have to be grounded upon reasons as to why various specific moral principles ought to be affirmed. Providing these reasons is the task of a more theoretical inquiry, primarily normative in nature. But it will also involve clarification of ideas about safety, loyalty to companies, and professional freedom and autonomy. Hence it will involve conceptual inquiries.

Descriptive inquiries, moreover, enter the case yet again when we attempt to discover the realistic options open to the engineer, as well as to provide some estimate of their likely consequences. Those inquiries will help us understand the business, social, and political realities in which the company operates. Only with this knowledge can we make a sound recommendation about what the engineer should do.

Study Questions

The following three illustrations are based upon actual cases. With respect to each, first state what you see as the normative or evaluative issues involved. Then identify any descriptive inquiries which you think might be needed in making a reliable judgment about the case. Finally, are there any ideas or concepts involved in dealing with the moral issues that it would be useful to clarify?

1 A County Engineer in Virginia demanded a 25 percent kickback in secret payments for highway work contracts he issued. In 1967 he made such an offer to Allan Kammerer, a 32-year-old civil engineer who was vice president of a young and struggling consulting firm greatly in need of the work. Kammerer discussed

the offer with others in the firm, who told him it was his decision to make. Finally Kammerer agreed to the deal, citing as a main reason his concern for getting sufficient work to retain his current employees (Fairweather, 54–5).

2 In 1970 Carl Houston was assigned as a welding superintendent to a nuclear power plant under construction. According to Houston, even his preliminary observations revealed numerous poor welds resulting from the use of poor procedures and improper materials. He considered this to be the result of improper training given to the welders and told his manager that if the situation was not corrected he would write to the main headquarters. Houston then described the situation to some of the subcontractors of the project, an act which led to his being fired for insubordination (Houston, 25–30).

3 The principal project finance officer for a large firm learned that an engineer had been charging his personal long-distance phone calls to a client for several months. The officer informed the engineer's supervisor about this. The charges were around $50 a month. It was extremely unlikely that the client would learn of them since the contract involved several million dollars, and if he did it would be easy to explain the situation as a bookkeeping error. The engineer was one of the firm's most valued employees, someone the supervisor regarded as the key to much of the hoped-for success of the firm during the next 2 years. The supervisor decided to ignore the situation. (From an unpublished case study written up by William Litle.)

SENSES OF "ENGINEERING ETHICS"

The word "ethics" (like the word "morals") has several distinct although related meanings. Corresponding to them are various senses of the expression "engineering ethics."

First, in the main sense used so far and in what follows, ethics is a discipline or area of study dealing with moral problems. Engineering ethics, accordingly, is the discipline or study of the moral issues arising in and surrounding engineering. As we have just seen, it involves normative (evaluative) inquiries, conceptual (meaning) inquiries, and descriptive (factual) inquiries into these moral issues. As understood here, the normative inquiries are central, with the conceptual and factual inquiries entering where relevant to support the normative inquiries.

Second, when we speak of ethical problems, issues, and controversies, we mean to distinguish them from nonethical or nonmoral problems. Here the word "ethical" marks a contrast between moral questions and questions of a political, legal, and artistic nature. Engineering ethics in this sense would refer to the set of specifically moral problems and issues concerning engineering.

Third, sometimes the word "ethics" is used to refer to the particular set of beliefs, attitudes, and habits which a person or group displays in the area of morality. Thus we say that Mussolini's ethics were different from those of Dwight Eisenhower, and both differed from those of Karl Marx. We also speak of the Romans' ethics, the Victorian ethic, the Protestant work ethic, and socialist ethics. In doing so we are referring to people's actual outlooks on moral issues. This sense is linked directly to the original sense of the Greek word *ethos,* which meant customs (as did *mores,* the Latin root of "morals"). As such, engineering ethics would be the currently accepted standards endorsed by various groups of engineers and engineering societies. The discipline of engineering ethics has the task of examining these accepted conventions to see if they are clear, justified, and sufficiently comprehensive.

This third sense of "engineering ethics" is purely descriptive because it concerns merely what engineers and others believe as regards moral problems in engineering. It is a usage in which social scientists especially might be interested, since they seek to describe and explain beliefs and actions related to morality. We will avoid it, however, because our main interest is in normative questions about correct beliefs.

Fourth, the word "ethics" and its grammatical variants can be used as synonyms for "morally correct." For example, people's actions and principles of conduct can be spoken of as either ethical (right, good, or permissible) or unethical (immoral), and individuals can be evaluated as ethical (decent, having moral integrity) or unethical (unscrupulous). In this usage, engineering ethics would amount to the set of justified moral principles of obligation, rights, and ideals which ought to be endorsed by those engaged in engineering. Discovering such principles and applying them to concrete situations is the central goal of the discipline of engineering ethics.

In order to simplify matters and avoid confusion, we shall restrict the expression "engineering ethics" to the sense in which it names a discipline. Nevertheless, we shall occasionally use the word "ethical" in its other senses when we speak of ethical problems, an individual's or group's ethics, and ethical conduct. The context will make it clear which of these is meant.

Study Questions

1 In response to a request by the *Monthly Newsletter* of The National Project on Philosophy and Engineering Ethics for brief definitions of engineering ethics, the following three suggestions were submitted and printed in the May/June 1979 issue. Which, if any, of the four senses distinguished above does each author seem to have in mind?

  a "Moral principles accepted by the profession relative to the practice of engineering." (Don Wilson, Engineer at Michael Baker, Jr., Inc.)

  b "The rights and responsibilities of those persons who practice the profession of engineering." (Al Flores, Department of Philosophy, California State University, Fullerton)

  c "If the task of *defining* engineering ethics is supposed to result in the articulation of a set of ethical principles which are individually necessary and jointly sufficient to distinguish engineering ethics from, say, business ethics, then there seems little hope that such a result is possible. For there will be an *engineering* ethics in this sense, that is, an ethics with principles which are engineering-specific, if and only if there are ethical aspects of engineering practice which are *unique* to the profession. . . . The ethical issues which bother the reflective engineer today revolve around questions of truth-telling, confidentiality, and obligations to employer/client vs. obligations to the public. But these ethical issues have been addressed by physicians for hundreds of years. . . .'' (Thomas A. Long, Department of Philosophy, University of Cincinnati, used with permission)

2 A full definition of engineering ethics will be linked with a definition of engineering. One wide definition of engineering says it is the application of science in the use of natural resources for the benefit of society or humanity. Such a notion would be presumptuous if it implied that engineers are the only ones who participate in engineering as so defined (Ramo, 12). But if this implication is avoided, the definition has the advantage of emphasizing how all of us—citizens, engineers, management, scientists, technicians, government regulators, etc.—are participants in engineering and as such should be concerned with the field of engineering ethics.

Consult a dictionary, an encyclopedia, and an introductory engineering textbook to see how they define engineering. Then carefully state the definition you prefer in connection with thinking about engineering ethics.

ENGINEERING ETHICS AND PHILOSOPHY

Much of engineering ethics can be viewed as part of applied philosophical ethics. Like medical, legal, and business ethics,† it is focused upon practical moral problems but seeks where possible to apply methods and theories derived from more general philosophical principles.

Thus, applied ethics necessarily makes contact in many places with ethical theory. This is because applied ethics is concerned with uncovering cogent moral reasons for beliefs and actions, as opposed to accepting uncritically whatever beliefs or actions might happen to strike one's fancy as being correct at a given moment. And the general moral principles to which such reasons make reference either explicitly or implicitly are directly linked to ethical theory.

We say, for example, that a given action is wrong because it amounts to accepting a bribe in the context of negotiating a contract, and bribes are

---

†We might note here that the closest cousin of engineering ethics is business ethics, which is also now undergoing rapid development. Since most engineers are salaried employees, and since engineering decisions are usually tied to business decisions, it is not surprising that moral problems in engineering and business often overlap. The question, for example, of what is wrong with offering and accepting more than nominal gifts when negotiating contracts for engineering services is at once an issue in engineering ethics and business ethics.

wrong because they unfairly influence judgment and decisions. This claim is based on a general principle that contract negotiations ought to be impartially centered on the merits of the contract alone. And that general principle, in turn, will be justified by appealing to a higher-order principle to the effect that we ought to be fair and impartial in certain situations. Such higher-order principles, when developed and integrated within theories of justice, constitute the broad philosophical perspectives on conduct that ethical theory can provide.

An analogy might help clarify the relationship between applied and general philosophical ethics. Engineering is itself an applied science which draws upon principles from general physics and mathematics. It is applied in the sense that it is aimed at finding practical solutions to concrete problems by applying either theoretical knowledge or the more intuitive understanding of a field gained after years of working within it. Pure science and mathematics, by contrast, are directed toward obtaining new knowledge, usually of a general sort. Practical engineering solutions are often obtainable without having to await final proofs of theoretical principles. On occasion particular engineering solutions will have a wide importance, however, and will provide insights that can be fed back into the more theoretical levels of science. Thus there arises a dynamic interplay between applied and so-called "pure" science.

Much the same can be said of the relationship between applied and general ethics, remembering that here the concern is with moral instead of scientific issues. General ethics tends to emphasize theoretical knowledge and examines practical cases only in order to illustrate and test theories. Applied ethics, by contrast, focuses upon concrete problems for their own sake, and invokes general theory where helpful in dealing with those problems. It is not the task of applied ethics per se to resolve long-standing philosophical disputes over the validity of general moral perspectives.

On occasion, however, applied ethics may well yield insights having a more theoretical importance. Future developments in engineering ethics seem likely to provide fruitful directions for thinking about enduring philosophical questions concerning the nature of collective responsibility, legitimate authority, justice within economic systems, and moral rights within complex institutions.

Philosophy for its part plays a major role in supplying the basic concepts and theories needed to clarify the nature of ethical problems and issues in engineering. Sometimes it can yield a heightened perception of those problems by placing them in perspective as aspects or instances of broader problems. For example, we shall see later how some appeals to codes of engineering ethics when trying to answer moral questions are problematic because the appeals presuppose a narrow kind of conventionalism in ethics. At other times philosophy can clarify general concepts which arise in thinking about any moral issue. Two examples of this to be dealt with in Chap. 2 are the concepts of ethical dilemmas and ethical relativism.

Historical Note

In many ways engineering ethics is still in its infancy. As a contrast, consider the current state of medical ethics. During the past two decades the moral problems in medicine have been given intensive interdisciplinary scrutiny by medical professionals, philosophers, psychologists, social critics, and lawyers. The results have been phenomenal. Currently there is a wide-spread interest in ethics among medical practitioners, and courses on medical ethics exist at most colleges and universities. Several new journals and an encyclopedia of biomedical ethics have been created, and each year numerous scholarly treatises on the subject appear.

By contrast, the large-scale interdisciplinary study of moral problems in engineering by engineers, philosophers, lawyers, managers, and technology theorists began only a few years ago. Centers for studies of these problems were only recently established at universities like Rensselaer Polytechnic Institute and the Illinois Institute of Technology. The first interdisciplinary conference took place in 1980, as did the publication of a scholarly bibliography on engineering ethics (Ladenson, 1980). And the first journal to emphasize philosophical articles on engineering ethics, *Business and Professional Ethics Journal,* was created in 1981.

This late development is ironic. Engineering is the largest profession numerically and affects all of us, often in large groups, in most areas of our lives. The skill of a surgeon's hand affects one patient at a time; the judgment of a design engineer can influence hundreds of lives at once. Medicine manifests itself in the yearly checkup, the miraculous cure, and the contents of the household medicine chest; but engineering products confront us virtually everywhere we turn our eyes and every time we do something.

Cognizant of their social responsibilities, engineers early on formed societies for the advancement of technical knowledge, for the setting of standards, and for the issuance of codes of ethics. Several societies have for years published case studies on possible violations of their codes. Yet the connection between engineering and philosophy is only now coming to be seen as the vital link it is.

WHY STUDY ETHICS?

What is the point of studying engineering ethics—or any applied ethics for that matter? And what can one expect from taking a course, or segment of a course, devoted to engineering ethics?

Of course there are many possible topics, and each of us will have special interests in probing different issues. Yet we believe that a shared practical goal in studying ethics should be to think clearly and critically about moral issues. To invoke a term widely used in ethics, the unifying goal should be to increase one's *moral autonomy* in developing and expressing reasoned moral views.

"Autonomy" literally means "self-ruling" or "independent." But not just any kind of independent reflection about ethics amounts to moral autonomy. Moral autonomy can be viewed as the ability to think, and the habit of thinking, rationally about ethical issues on the basis of moral concern. This foundation of moral concern, or general responsiveness to moral values, derives primarily from the training we receive as children in sensitivity to and consideration of the needs and rights of others. Where such training is absent, as it often is with abused or neglected children, the tragic result can be an adult sociopath capable of murdering without compunction. Sociopaths, who by definition lack a sense of moral concern and guilt, are never morally autonomous—no matter how "independent" their intellectual reasoning about ethics may be.

Adult moral concern, of course, can be evoked (or numbed) on specific occasions by any number of influences: friends, ministers, social events, novels, movies, and inspiring teachers of whatever subjects. It would be surprising if the topics dealt with in ethics courses did not call forth moral concern. Nevertheless, the main point of taking a course on applied ethics should be to improve the ability to reflect critically on moral issues.

This can be accomplished by improving various practical skills that will help produce effective independent thought about moral issues. As related to engineering ethics these skills include the following:

1 Proficiency in recognizing moral problems and issues in engineering. This involves being able to distinguish them from, as well as relate them to, problems in law, economics, religious doctrine, or the descriptions of physical systems.

2 Skill in comprehending, clarifying, and critically assessing arguments on opposing sides of moral issues.

3 The ability to form consistent and comprehensive viewpoints based upon consideration of relevant facts.

4 Imaginative awareness of alternative responses to the issues and creative solutions for practical difficulties.

5 Sensitivity to genuine difficulties and subtleties. This includes a willingness to undergo and tolerate some uncertainty in making troublesome moral judgments or decisions.

6 The learning of a common ethical language, which is necessary in order to know how to express and defend one's moral views adequately to others.

The exercise of moral autonomy carries no guarantee that everyone will arrive at either the truth or the same verdicts. There do seem to be some basic moral differences which are not reducible to disagreements over facts or errors in logical inference. Perhaps this is inevitable in a subject like morality which, as Aristotle noted long ago, is not as precise and clear-cut as arithmetic. It is the beginning of tolerance to allow considerable room for disagreement among autonomous, reasonable, and responsible moral agents.

Obviously, then, the aim of teaching engineering ethics should not be to produce a unanimous conformity of outlook among engineers even though such conformity could be achieved by resorting to indoctrination, authoritarian and dogmatic teaching, hypnotism, drugs, or other autonomy-destroying techniques. Indeed, if one major goal of engineering ethics should be to uncover ways of promoting tolerance in the exercise of moral autonomy by engineers, the same goal should be present within courses on engineering ethics.

Most of us value moral autonomy for its own sake. Its exercise is central to what we think of as the possession of a mature moral outlook, that is, one which is something more than second-hand and passively adopted. Yet we also value it because we believe that it tends to lead to morally responsible conduct or is an integral part of being a responsible person. Certainly some faith in the ability of people to reflect on moral issues as morally autonomous individuals is presupposed in any attempt to engage in serious dialogue about ethics. And an affirmation of the individual's right to develop and exercise reasoned moral perspectives presupposes that most people have a large capacity for acting responsibly out of humane values. It is in this spirit that we, as authors, wish to enter into the discussion of engineering ethics.

Study Questions

1 Most of us agree that the dogmatic teaching of ethics aimed at indoctrination can threaten the exercise of moral autonomy. Does this mean that college teachers should withhold expression of their own views on moral issues?

2 Describe a clear-cut example of where one person tries to indoctrinate another into holding a particular value or belief. Does the attempt center more on what is said (for example, false views, racist views, one-sided views, etc.) or on how the view is presented (for example, with intimidation, intolerance, the suppression of criticism, etc.)?

3 Respond to the following argument: "The primary goal of a course on engineering ethics ought to be to have students master the standards of professional conduct

specified in the major engineering codes of ethics. This is because the codes are formulated by the engineering professional societies which officially speak for the engineering profession on moral issues."

SUMMARY

Various things may be meant by the expression "engineering ethics." In this book *engineering ethics* will mean the examination of the moral issues in engineering and the field of study which results from that examination. It centers on a *normative* or evaluative inquiry into how people and organizations involved in engineering ought to behave, and what laws, codes of ethics, and institutional norms morally ought to be in effect. But it also involves *conceptual* inquiries aimed at clarifying the meanings of key ideas and issues and *descriptive* or factual inquiries designed to provide relevant factual information.

The issues in engineering ethics are wider in scope than the moral problems confronted specifically by engineers. For they include the moral problems bearing on engineering faced by many other people, including consumers, managers, scientists, technical writers, lawyers, and government officials. Even though the emphasis in this book will be on those problems as they relate to engineers, it will become clear in the course of our discussion that the responsibilities of engineers must be understood in conjunction with the rights and obligations of these other people, especially consumers and managers.

Engineering ethics is intimately tied to philosophical ethics, and it can be viewed as a branch of applied philosophical ethics. This is especially true wherever general ethical theories and distinctions are applied in order to (1) enrich our understanding of the nature of the moral problems and issues in engineering, (2) provide reasoned responses to those problems, and (3) clarify the meanings of key concepts and distinctions.

Different courses in engineering ethics will focus on different particular topics, and each student will come to such courses with special interests. Nevertheless, the unifying practical aim in studying and teaching engineering ethics is to help foster moral autonomy. *Moral autonomy* is the ability to arrive at reasoned moral views based on the responsiveness to humane values most of us were taught as children. This does not require that a person always reach the "correct" moral view. But it involves displaying competencies and sensitivities such as the following: the abilities to discern moral problems and to clarify them, to work out reasoned and sometimes creative responses to them while taking account of opposing viewpoints, and to exercise the verbal and communicative skills relevant to discussing one's views with others.

CHAPTER 2

MORAL REASONING

On October 10, 1973, Spiro T. Agnew resigned as Vice President of the United States amidst charges of bribery and tax evasion related to his previous service as County Executive of Baltimore County. As County Executive during 1962 to 1966 he had the authority to award contracts for public-works projects to engineering firms. In exercising that authority he functioned at the top of a lucrative kickback scheme (Cohen).

Lester Matz and John Childs were two of the many engineers who participated in that scheme. Their consulting firm was given special consideration in receiving contracts for public-works projects so long as they made secretive payments to Agnew of 5 percent of fees from clients. Even though their firm was doing reasonably well, they entered into the arrangement in order to expand their business. They felt that in the past they had been denied contracts from the county because of their lack of political connections.

Types of Moral Problems

If we say that Matz and Childs confronted a moral problem, we might mean several things. We could be calling attention to the fact that they were in a situation calling for a decision involving moral considerations: that is, considerations about fair practices and avoiding deception. Or we could mean they were tempted to ignore these moral considerations and violate moral rules like "Do not cheat." But we would probably not intend to imply there was any serious doubt or difficulty in determining what they were morally required to do—that was clear-cut. The only uncertainty was whether they would decide to do what was morally proper.

Usually, however, when we speak of moral problems we have in mind situations where what ought to be done is not so straightforward and obvious. Those situations may involve two sorts of murkiness or complexity which set them in sharp contrast to the case of Agnew, Matz, and Childs.

First, it may be unclear to the individuals involved which, if any, moral considerations or principles apply to their situation. An engineer starting a new job, for example, may have doubts about whether it is morally permissible to accept an expensive desk set as a gift from a salesperson with whom his company does business. Would this be accepting a bribe? Would it create a conflict of interest? Perhaps a conversation with a colleague will answer these questions. But there will always be troublesome cases where there is considerable vagueness about whether the "gift" is an innocent amenity or an unacceptable bribe.

Second, it may be perfectly clear which moral principles apply to one's situation. The difficulty instead might be that two different moral principles, both of which apply to one's situation, come into conflict or that one principle seems to point simultaneously in two different directions. These kinds of moral problems, called *moral dilemmas*, deserve special attention because of the practical difficulties they sometimes pose, and also because of what they reveal about the complexities of morality.

MORAL DILEMMAS

Most moral dilemmas are situations where a person confronts two or more moral obligations, duties, rights, or ideals which have come into conflict with one another, and at least on the surface not all of them can be fulfilled or respected. But it is also possible for one moral principle to have two or more incompatible applications in a given situation.

Such conflicts occur frequently, although usually there is only moderate difficulty in seeing what should be done. We make a promise to a friend, thereby creating an obligation to do what we have promised. Our parents become ill and staying home to help them prevents us from keeping the promise. The dilemma, which consists of a conflict between the duty to keep promises and an obligation to one's parents, can usually be resolved by an apologetic phone call to the friend. Or again, we make one promise to our employer and another to a colleague, and it turns out that we cannot keep both. Here the general duty to keep promises has two incompatible applications. Once again, an apology to the offended party will often settle the matter.

Yet dilemmas are not always so easily dealt with. Resolving some of them can require searching, even agonizing, reflection. Contemporary engineering practice makes it virtually inevitable that nearly all engineers will be confronted with some moral dilemmas of this sort during their careers. Let us consider several examples, beginning with some vivid illustrations drawn from works of fiction.

Obeying Laws; Protecting Innocent Life

The events in Herman Melville's novel *Billy Budd* illustrate a conflict between personal and professional loyalties. While at sea on a military vessel, Billy Budd is falsely and maliciously accused by the master-at-arms of encouraging insubordination and possible mutiny among the crew. Completely shocked and dumfounded, Budd spontaneously strikes the man, unintentionally killing him. The shipmaster, Captain Vere, must decide whether to recommend the seaman's execution to a drumhead court. As a military officer, Vere believes his duty to the state requires the execution. But his personal view, having witnessed the event and knowing of the cruelty of the master-at-arms, is that Billy is morally blameless, and certainly morally innocent of a capital offense.

Vere's final decision is that his private conscience must yield to stern military duty, and he argues as follows before the court:

> In receiving our commissions, we in the most important regards ceased to be natural free agents. When war is declared are we, the commissioned fighters, previously consulted? We fight at command. If our judgments approve the war, that is but coincidence. So in other particulars. So now. For suppose condemnation to follow these present proceedings. Would it be so much we ourselves that would condemn as it would be martial law operating through us? For that law and the rigor of it, we are not responsible. Our vowed responsibility is in this: That however pitilessly that law may operate, we nevertheless adhere to it and administer it (Melville, 68–69).

While this first example is not from engineering, it raises relevant issues concerning obeying laws and orders from superiors. It suggests parallels between moral dilemmas in the military and those arising within some bureaucracies where there are strong authority relationships.

Historically there is a link between engineering and the military. While advances were occurring in many fields of emerging technology, from mining to irrigation, rulers throughout the centuries were particularly interested in fortifications and machines of war. The artisans who designed and built these ingenious military devices were called "engineers." The earliest engineering schools in a number of countries, including France and the United States, were military schools. The term "civil engineer" was coined later to differentiate between military and nonmilitary end-uses of engineering.

Loyalties to Two Groups

A situation involving an engineer within a war context is provided by Pierre Boulle's *The Bridge over the River Kwai*. The novel is about someone who by following rules narrowly comes to lose sight of the wider context of the work he is doing.

During World War II the Japanese force a group of Allied prisoners to construct a bridge for a Burma-Siam railroad intended to greatly speed the Japanese war effort. Among the prisoners is an engineer, Captain Reeves, who is dismayed by the shoddy workmanship of the Japanese engineer directing the project. Reeves complains bitterly to his commanding officer, Colonel Nicholson, about the bridge's location in a swamp lacking a solid foundation, about the use of soft timber in its construction, and about the lack of a master plan for a structure meant to stretch 100 yards. Reeves further protests that the bridge might easily collapse the first time a train crosses it.

The dilemma confronting Reeves and Nicholson arises when the prisoners come to feel their pride, dignity, and sanity can be best preserved through the disciplined construction of a safe and pleasing bridge. In diligently pursuing this goal, they take charge of the entire project and turn the bridge into something of a masterpiece. But as they do so, they forget their obligation to benefit the Allies by allowing a faulty bridge to be built. Hence the dilemma consists of Reeves' and Nicholson's obligation to support their men in building a good bridge as opposed to their obligation to support the Allies' war effort by building a shoddy bridge of no help to the enemy. It involves a conflict between loyalties to two groups. The resolution of the dilemma has Colonel Nicholson dying in defense of the bridge against Allied attempts to blow it up.

Duties to Employers and the Public

A more recent fictionalized dilemma is presented in the movie *The China Syndrome* and the novel based on it by Burton Wohl. The main plot begins when a tremor shakes the nuclear facility where Jack Godell (Jack Lemmon in the movie) is the shift manager. At first Godell dismisses the tremor as the result of a stuck valve which has affected a turbine. Subsequent testing, however, leads to a small shudder in a pump and later to the discovery of a small leak of radioactive cooling water. In checking the files Godell uncovers omissions and forgeries of required x-rays of the welds on water pipes used in cooling the reactor core.

Godell concludes that the plant is unsafe and should temporarily be shut down until new x-rays are made. Faulty welds might lead to pipe failures and inadequate cooling of the reactor core. In turn, this could cause a meltdown in which the fuel might burn through its container into the earth below "headed toward China." But first it would hit the local water table and erupt into a cloud of radioactive vapor, enveloping hundreds of thousands of people.

The chance of this happening is small, but Godell is alarmed. He feels a strong loyalty to his superiors and wants to work through normal channels and to support the integrity of the plant's operations. Yet he is unable to convince his supervisor and colleagues of the danger. His dilemma involves competing obligations to obey and to disobey company procedures. It can also be formulated in terms of the competing rights of the public to safety and of his employer to professional discretion. Fearing that any further vibration could cause a disaster, Godell seizes control of the plant and demands that he be allowed to broadcast his views over television.

Duties to Clients and the Public

As a last example from fiction, we will examine William Golding's novel *The Spire,* a rich allegory about fidelity, creativity, and the way self-deception may warp concern for safety within engineering. Set in England during the middle ages, the plot revolves around the construction of a 400-foot spire atop an aging cathedral. Success in the project would mean developing technology well beyond its then current state, and this provides the motivation for the master builder commissioned to undertake the project.

Yet as the master builder assesses the weaknesses of the foundations supporting the church, he is led to suspect that the stone and glass spire cannot be supported properly. Thereafter his suspicions are repeatedly confirmed to the point where his best professional judgment—what he humbly calls his "guesses"—indicate the task is both futile and dangerous. Nevertheless, the priest who is his client desperately clings to a vision of the spire as a "prayer in stone" and urges the craftsman on.

Both of the conflicting loyalties depicted in this story arise directly out of the professional duties of engineers. One concerns the basic safety of a project. The other involves a combination of commitment to the client and excitement at pushing technology past its current boundaries. The dilemma has to do with the clash between competing obligations to ensure the safety of eventual users of the church (including in this case the client) and to meet the client's expressed needs. It also involves a clash between the ideal of pursuing new technology and the rights of people not to be unknowingly subjected to dangers.

The master builder gradually becomes biased as he allows his excitement over the project and the personal influence of the priest to lead him into a disregard for safety. As the story ends, the entire structure is slowly crumbling.

A DC-10 Case: Duties to Employers and the Public

Let us turn now to a tragic nonfictional dilemma of recent history, one documented in the book *Destination Disaster* (Eddy). In 1974 the first crash of a fully loaded DC-10 jumbo jet occurred over the suburbs of Paris, killing 346 people. Apart from the record number of people killed, one thing distinguished the incident: It was known in advance that something like it was bound to occur due to defective design.

The fuselage of the plane was developed by Convair, a subcontractor for McDonnell-Douglas. Two years earlier Convair's senior engineer directing the project, Dan Applegate, had written a memo to the vice president of the company itemizing the dangers that could result from the design. He accurately detailed several ways the cargo doors could burst open during flight, depressurize the cargo space, and thereby collapse the floor of the passenger cabin above. Since control lines ran along the cabin floor, this would mean a loss of control of the plane. Applegate recommended redesigning the doors and strengthening the cabin floor. Without such changes, he stated, it was inevitable that some DC-10 cargo doors would open in midair, resulting in crashes.

In responding to this memo, top management at Convair did not dispute either the technical facts cited by Applegate or his predictions. Company officers maintained, however, that at that time of intense competition (Newhouse), the possible financial liabilities Convair might incur prohibited them from passing on this information to McDonnell-Douglas.

It might be argued that as a loyal employee Applegate had an obligation to follow company directives, at least responsible ones. Perhaps he also had family and other personal obligations which made it important for him not to jeopardize his job. Yet as an engineer he was obligated to protect the safety of those who would use or be affected by the products he designed. Thus the dilemma he confronted involved a clash between at least two general professional obligations, and possibly a clash between professional and personal obligations as well.

While this is an extreme case, it nevertheless illustrates a common class of moral dilemmas in engineering. Given that the vast majority of engineers are salaried employees, it is very likely that employer-based duties will on occasion conflict with public-based duties. For this reason we will take all of Chap. 5 to examine the relations between these kinds of obligations. Elsewhere we will look at nonfictional examples of other sorts of dilemmas involving obligations to clients and to employers, to clients and to the public, to colleagues and to one's company, to competing public sectors, and so on.

Resolving Dilemmas

Often there are ways of resolving moral dilemmas successfully, as we mentioned earlier. Moral insight and imagination, aided by good advice from others, enable us to uncover ways of reconciling what seem to be unreconcilable obligations. Yet in some instances no happy resolution of a dilemma may be possible. When that happens, what of the obligation which cannot be met? Does it automatically cancel out and vanish? Is the person involved perhaps relieved of the responsibility to meet it?

Notice that even when we break a promise to someone through no fault of our own in a situation where we cannot inform the person in advance, we feel we have wronged that individual and must somehow make it up to him or her. We seek positive ways to make amends: We explain the situation to the person, try to restore mutual understanding and trust, provide assurances of continuing concern by remedying any harm resulting from the broken promise. Perhaps in cases where unintentionally broken promises produce serious consequences, where no means exist to make amends fully, we must live with some remorse and be more creative in finding ways to deal with our guilt.

Positive steps can and should be taken to keep many dilemmas from arising, or at any rate to keep them from resulting in tragic conclusions. And engineers, management, and others are struggling to develop ways at least to remove the isolation in which many engineers have had to confront dilemmas in the past. But in the next section we will introduce several types of ethical theories which can aid in sorting out, understanding, and dealing with moral dilemmas when they do arise.

Study Questions

1 Locate and read a complete work of fiction related to engineering which involves a moral dilemma. Then write an essay in which you (a) discuss in detail the origin and nature of the dilemma, and (b) explain whether or not you agree with how the person involved resolved the dilemma, giving reasons for your opinion.

You might consider the following works: Pierre Boulle, *The Bridge over the River Kwai;* Eugene Burdick and Harvey Wheeler, *Fail-Safe;* Graham Greene, *A Burnt-Out Case;* William Golding, *The Spire;* Henrik Ibsen, *The Master Builder;*

John D. MacDonald, *Condominium;* Louis V. McIntyre and Marion B. McIntyre, *Scientists and Engineers: The Professionals Who Are Not;* Nevil Shute, *No Highway;* Kurt Vonnegut, Jr., *Player Piano;* Burton Wohl, *The China Syndrome.* Additional possibilities are given in Samuel C. Florman's *The Existential Pleasures of Engineering.*

2 Follow the directions for question 1, except this time select your example from a current newspaper, magazine, or journal.

3 Interview at least one person involved in engineering, preferably an engineer. Ask them to describe one or more moral problems they have confronted. Is the problem they describe a moral dilemma? If so, explain which competing moral obligations, ideals, or principles caused it and how you might have resolved it.

4 According to Kenneth Kipnis, a professor of philosophy, the design engineers share the blame for the death of the passengers in the DC-10 crash described above. Kipnis contends that the engineers' overriding obligation was to obey the following principle: "Engineers shall not participate in projects that degrade ambient levels of public safety unless information concerning those degradations is made generally available" (Kipnis, 1981, 82). Do you agree or disagree with Kipnis, and why?

THEORIES OF RIGHT ACTION

What makes actions morally right or morally wrong? And what are the most basic reasons why we morally ought or ought not to act in various ways? Perhaps there are no simple general answers to these questions. Yet philosophers have attempted to provide answers, and their efforts have often led to helpful systematic theories about morality. Those ethical theories fall into three categories, depending upon the relative importance assigned to the consequences of actions (or to goals), to duties, and to rights (Dworkin, 169–173).

In order to work with a concrete illustration, return to the kickback scheme described at the beginning of this chapter. We assumed it was clear that the actions of the participants were unethical. But what reasons can be given to support this assumption? *Why* was it wrong for the engineers to make secretive payments to Spiro Agnew in return for being given preference in the awarding of contracts for public projects?

One answer is that bad consequences resulted. Other engineering firms were harmed by not having a chance to obtain the contracts they may have been best qualified to receive. The system also removed the potential benefits of healthy competition among a wider range of firms, benefits such as lower costs to the public. Perhaps more important, discovery of the scheme led to a loss of trust in public officials, a trust important for the well-functioning of government. And the individuals themselves eventually suffered greatly after they were caught.

Goal-based theories assert that the goals of producing good consequences and avoiding bad ones are ultimately the only morally important considerations. Actions are right or wrong solely by virtue of their good or bad effects. Because goal-based theories emphasize exclusively the ends served by actions, they are also called "teleological theories" from the Greek word *telos,* meaning "end" or "purpose."

A different answer to what was wrong with engaging in the kickback scheme would have us focus on the inherent nature of the act itself, rather than merely its consequences. The actions involved were secretive and intended to keep outsiders deceived about what was going on. They were also inherently unfair to others who were denied equality of opportunity to bid for the contracts. Hence the actions, irrespective of their actual consequences, violated at least two basic principles of duty: "Avoid deceiving others" and "Be fair."

Duty-based theories thus assert there are duties to perform and certain types of acts to avoid, even though doing so may not always produce the most good. Actions are right when they conform to principles of duty which encompass more than good consequences, and wrong when they violate those principles. For example, one is always duty-bound to be fair and not to deceive, even when such practices will not lead to the best consequences in given situations. Because they emphasize duty, the Greek word for which is *deon,* these theories are also called "deontological theories."

Yet a third answer to why it was wrong to participate in the kickback scheme is that it violated the rights of other people. A shared understanding exists that there will be equality of opportunity in seeking public contracts and that elected officials will grant contracts based on merit, not bribes. Against this background, qualified persons or firms acquire the right to unbiased consideration of their contract proposals, and these rights were violated by the kickback scheme. It might also be argued that the public's rights to responsible behavior from their representatives were violated.

Thus *rights-based theories* treat actions as wrong because they violate moral rights. Like duty-based theories, they deny that good consequences are the only or even the primary moral consideration. But rights-based theories say we have duties to other people because people have rights which ought to be respected. By contrast, duty-based theories tend to treat rights as mere secondary correlates of more fundamental duties. For example, they recognize a person's right to fair treatment solely because people in general have a duty to be fair to others.

All three types of theories give some importance to duties, rights, and the goals of producing good and avoiding bad consequences. They differ, however, in the relative importance and prominence they assign to those factors, as the table below summarizes.

Let us look at examples of influential versions of these three types of theories.

	Basic	**Secondary**
Goal-based theories	Good consequences	Duties, rights
Duty-based theories	Duties	Rights, good consequences
Rights-based theories	Rights	Duties, good consequences

A Goal-Based Theory: Utilitarianism

Utilitarianism is by far the most influential goal-based theory. It holds that all sound moral reasons reduce to one fundamental principle: We ought always to produce the most good for the most people. This theory is goal-based in that it treats good consequences as all that matter in determining what is right or wrong. It is a special type of goal-based theory in that it insists the good of *everyone affected* must be taken into account.

Actually there are two main versions of utilitarianism which differ in the roles they assign to everyday moral rules (Brandt, 380–405).

Act-utilitarianism says we should examine each individual *action* to determine whether it will produce the most good for the most people. General rules like "Keep your promises" and "Don't make bribes" are viewed as rough rules of thumb which summarize what is usually the best thing to do. But such rules can have many exceptions as applied within concrete situations.

Rule-utilitarianism, by contrast, makes *rules* primary: We ought always to act on those rules which if generally followed would produce the most good for the most people. Individual actions are justified only if they conform to such rules. Thus a rule-utilitarian might say we ought to keep a promise or not offer bribes, even though such acts do not always have the best consequences, because the general practice leads to good consequences.

Both versions of utilitarianism, as with all goal-based theories, regard the criterion for correct action to be the goodness produced, as is true of all goal-based theories. But what is goodness? Different versions of utilitarianism provide different answers to this question, depending on what is seen as *intrinsically good.* An intrinsic good is something which is good in and of itself, rather than merely useful as a means to obtaining other good things. Most of us view eating a delicious meal as intrinsically good and painful surgery as intrinsically bad. But the surgery may be an *instrumental good:* an "instrument" or means to produce the intrinsic goods of health and the absence of pain.

Jeremy Bentham (1748–1832) said that only pleasures are intrinsically good. He added that the degree of goodness of pleasures is measured solely in terms of their quantities: for example, in terms of their intensity and duration. John Stuart Mill (1806–1873) said that happiness, involving a life of richly varied pleasures, is the only intrinsic good. He differed sharply with Bentham by insisting that the worth of pleasures varies by more than mere quantity—there is also quality. And for him the pleasures of intellectual endeavors, appreciation of art, and moral virtue were inherently higher in "quality" than any purely physical pleasures.

More recent utilitarians have viewed additional things as intrinsically good. G. E. Moore (1873–1958) counted friendship, love, and appreciation of art as inherently good, even though they might sometimes lead to unhappiness and pain.

Many economists have embraced utilitarianism, but have been dissatisfied with the traditional theories of goodness given by Bentham, Mill, and Moore. Those theories tend to identify goodness with inner psychological states which may be difficult to observe and measure—states such as pleasure, happiness, and knowledge. Instead, economists have emphasized actions expressing personal preferences. For them, something is good to a person to the degree that the person prefers it over other available options. According to this version of utilitarianism, the right decision is the one which produces the greatest satisfaction of the preferences of all the people affected.

Duty-Based Theories: Kant, Ross, and Rawls

Immanuel Kant (1724–1804) is the most famous of the ethicists who emphasized duties as more important than the goal of producing good consequences. In Kant's view to make a moral judgment is to categorically prescribe or command actions which can be willed universally.

The idea of a *"categorical* command" can best be explained by first considering nonmoral prescriptions or commands. Kant said these are always set forth with some special condition attached. If a mugger with a gun commands, "Your money or your life," he is understood to mean, *"If you want to avoid being shot, hand over your money."* Again, a doctor's order to stop overeating tacitly means, *"If you desire to become healthier, stop overeating."* Because such commands are made on the basis of an implied "if" clause which specifies a condition or "hypothesis," Kant labeled them "hypothetical imperatives." Naturally such prescriptions apply to us only if we have the relevant desire that would be satisfied by obeying the command.

Moral imperatives are different. According to Kant they have no conditions attached. They are unconditional or categorical imperatives which require us to do things whether we feel like it or not. For example, they prescribe without qualification that we keep our commitments, not lie, be honest, and protect human life. They do not make hypothetical commands like, "If you want to be happy, keep your promises."

Thus moral imperatives enjoin that we obey them out of an autonomous commitment to morality—that we choose to do the right thing simply because it is seen as the right thing. Kant emphasized moral autonomy to the point of saying that right actions are simply those which express autonomous commitments. In acting morally, we can view ourselves as the creators or authors of the moral principles we act upon.

For Kant, categorical imperatives are binding on us only if they are also applicable to everyone. That is, moral reasons and principles are those which we are *willing* to have all people act upon, and which, without inconsistency, we can *conceive,* of all people acting upon. In this sense they must be *universalizable*.

Most everyday moral rules pass this test. For example, we can imagine—and favor—having everyone obey the command "Keep promises." By contrast, we cannot imagine people obeying the command "Keep promises except when you don't feel like it," for if everyone did so, promises would no longer be possible. Whenever someone attempted to give their solemn word to us by uttering the words "I promise," we would merely laugh. Serious promises are understood as not being subject to the whims of those who feel now and then like obtaining special advantages for themselves. Thus we are not *willing* to encourage people to break promises whenever they find it convenient, and we are caught in a *contradiction* when we try to imagine a situation in which everyone acted that way.

Why was it so important for Kant that we act toward others according to principles that are universalizable? In Kant's view all persons have an inherent worth which must enter into our moral deliberations. He thought that acting on universalizable principles was the only way to ensure that we do not misuse others in a way that denies their worth. In his famous phrase, people should be viewed as "ends in themselves," that is, as beings to be treated as more than just means toward our own personal ends. To steal, kill, or rape is to reduce one's victims to mere means for gratifying one's desires, thereby disregarding their intrinsic moral worth and their equality with oneself.

Kant's ideas of autonomous commitment to categorical imperatives, universalizable moral principles, and treating people as ends in themselves have had a wide influence. Yet most ethicists reject Kant's view that everyday moral rules should never have exceptions. Given the complexities of the world, it is inevitable that simple rules such as "Keep promises" will occasionally come into conflict with other rules, generating moral dilemmas. When that happens, one of the rules might have to allow for an exception.

A recent philosopher especially sensitive to this fact was David Ross (1877–1971). Ross suggested that moral reasons are based upon a set of duties, and the list of those duties includes familiar entries that Kant would also have endorsed: keep one's commitments; make reparation when one harms others; show gratitude for kindness extended by others; be fair and just; be kind and beneficent; seek to improve one's intelligence and character; make special efforts not to harm others. But for Ross, unlike for Kant, each of these duties can have exceptions when, in a given situation, it comes into conflict with a duty which is more compelling. To underscore this fact, Ross introduced a technical expression which is now widely used. He called them *prima facie duties*. For Ross, it was intuitively obvious that each of the above duties is binding on us. But it is often very difficult to decide what we actually ought to do in situations where two or more prima facie duties come into conflict. No easy test, including Kant's universalizing test, can help us. We must simply be sure to reflect calmly and fully before arriving at a judgment, and always to remain open to any criticisms by others which might help us correct our judgments.

John Rawls, a leading contemporary ethicist, has sought to build upon Kant's ideas of autonomy and universality, while moving beyond Ross's heavy reliance on intuition. According to Rawls, valid ethical principles can be viewed as those which would be voluntarily agreed upon by all rational persons in an imaginary "contracting" situation. The persons in the hypothetical situation are characterized by several features: (1) They lack all specific knowledge about themselves: for example, their particular desires, intelligence, and achievements. This ensures they will not be biased by self-interest in their deliberations. (2) They do have general knowledge: for example, about human psychology, the economics and politics of society, and science. (3) They have a rational concern for promoting their long-term interests. (4) They seek to agree with each other about the principles they will voluntarily follow as a group. That is, they form a moral agreement or contract to abide by principles to which they all subscribe.

Rawls's view is Kantian in that it emphasizes the autonomy each person exercises in forming agreements with anyone else willing to reason about moral principles. And he believes that placing ourselves in this hypothetical contracting situation helps us to reason more easily about those principles because it allows us to check out our intuitions as we attempt to discern which are valid and which are not.

Rawls argues that there are two basic moral principles applicable to social institutions which all rational people would agree to in the hypothetical situation. They may be paraphrased as follows: (1) Each person is entitled to the most extensive amount of liberty compatible with an equal amount for others. (2) Differences in social power and economic benefits are justified only when they are likely to benefit everyone (Rawls, 60).

Rights-Based Theories: Locke and Melden

The third type of ethical theory makes moral rights fundamental. According to it, most duties and obligations arise because people have rights, not vice versa. For example, individuals do not have the right to live merely as the result of other people's duty not to kill them. Instead, possessing the right to live serves as the justification for why others ought not to kill.

John Locke (1632–1704) argued that to be a person entails having rights—human rights—to life, liberty, and the property generated by one's labor. His views had an enormous impact at the time of the French and American revolutions. The words in the Declaration of Independence are not far from his own:

> We hold these truths to be self-evident; that all men are created equal; that they are endowed by their creator with inherent and inalienable rights; that among these are life, liberty, and the pursuit of happiness.

Locke's own version of a human rights doctrine was highly individualistic. He viewed rights primarily as entitlements to keep others from trespassing upon and meddling in one's life. For him, rights were primarily the "negative" ones which place obligations on others not to interfere. This aspect of his outlook is represented on the contemporary political scene in the libertarian ideology, with its emphasis on protection of private property and the condemnation of welfare systems (Nozick).

Other rights-based theories, however, conceive of rights as intimately related to communities of people. A contemporary philosopher, A. I. Melden, has argued that having moral rights presupposes an individual has the capacity to show concern for others and to be accountable within a moral community (Melden, 1977). The extent of rights, in his view, always has to be determined in terms of interrelationships among people. Melden's account allows for more "positive" rights to community benefits needed for living a decent life. Thus it lays the groundwork for recognizing the welfare rights of the needy.

Not all moral rights are simple human rights. Many arise from special relationships and roles which people might have. A promise, for example, gives rise to the special right to have the promise kept. And many of the rights we will examine later arise within institutions and professions, such as the right of managers to direct a company and the rights of engineers to warn the public about unsafe technological products. These are institutional moral rights rather than human moral rights.

While rights-based ethicists treat rights as fundamental, they generally do not treat them as the sole moral criteria. Other considerations may arise that make it inappropriate to exercise a particular moral right. For example, although an individual has the right to give all his chocolates to one child to eat in front of other children, it might be inconsiderate to do so. Although at first it sounds paradoxical, having the right to perform an act does not automatically mean that the act is right to perform.

Uses of Ethical Theories

What should be expected of these ethical theories in dealing with concrete moral problems, and especially in dealing with moral dilemmas? We should not expect the theories to provide a simple set of rules for resolving all troublesome moral problems. There are no such moral algorithms.

The theories often do provide straightforward accounts of what makes various actions right or wrong. We saw an example of this with the kickback scheme. Thus the theories can be invoked to justify or give reasons in support of moral judgments. The kickback example also illustrated how different types of theories often yield the same practical conclusions about what is right or wrong. This should come as no surprise, since each of the theories grows out of our shared moral experience. The theories represent alternative attempts to identify comprehensively the structure and order in that experience.

An ordering and structuring of moral experience can be helpful in approaching moral dilemmas, even though it does not always yield easy solutions to the dilemmas. Theories may help a person identify the moral factors which, from that individual's value perspective, are the most important or significant to weigh before making a decision.

Theories also provide a quickened sense of why dilemmas arise. Certainly a view like Ross's, involving a set of equally basic duties, can help make vivid why conflicts among duties arise. Rights-based ethics also allow for a plurality of rights which might come into conflict. And while utilitarianism is based on just one basic moral principle, it allows for a wealth of good and bad consequences which may come into conflict.

A moral theory might be thought of as providing a general framework within which to approach the consideration of moral dilemmas. Resolving a dilemma involves adding detail to the framework by prompting a refinement of the theory to which we are committed. For it makes us recognize the greater importance of one duty over another, one right over another, or one type of good consequence over another within particular circumstances.

Rational Egoism

The sample theories we have examined all treat morality as requiring more than a consuming preoccupation with one's private self-interest. In effect, they differ only in the ways they interpret how we ought to take into account the interests of others. Thus it is no surprise that most of the ethicists mentioned endorsed some version of the Golden Rule: "Do unto others as you would have them do unto you"; or stated negatively, "Do not do unto others as you would not have them do unto you." All of the theories help explain why there is one standard question used to instill moral concern in children: "How would you like it if someone did that to you?"

This way of thinking about morality has not been without its critics who wish to emphasize the importance of self-interest. Rational egoism is the view that maximizing one's own good ought to be the ultimate goal of each of us. According to its proponents, moral values are reduced to prudential concern for oneself, but always a "rational" concern requiring consideration of one's long-term interests (Rand). This is a goal-based theory, since it emphasizes good consequences. But it differs sharply from utilitarianism which prescribes considering the good of all people equally.

Defenders of egoism sometimes expound two claims which bring their views somewhat closer to the preceding theories. First it is standard for them to draw a distinction between narrower and wider forms of self-interest. To be selfishly preoccupied with one's own private good to the point of indifference and disregard for the good of others will generally cut one off from rewarding friendships and love. Thus the "paradox of happiness": To seek happiness by blinding oneself to other people's happiness leads to one's own unhappiness. Personal well-being generally requires taking some wider interest in others, although the rational egoist insists that the only reason for showing an interest in others is for the sake of oneself.

Second, some egoists contend that an ironic consequence of everyone rationally pursuing their self-interest is that everyone benefits. For example, the classical economist Adam Smith and his contemporary defender Milton Friedman believe that society benefits most when (1) individuals pursue their private good, and (2) corporations (as expressions of many individual wills) pursue maximum profits in a competitive free market (Smith; Friedman, 1962). The idea is that this will make the economy prosper and thereby benefit everyone.

We will have occasion to return to the views of Smith and Friedman in Chap. 6. Here we express our strong doubt that private pursuit of self-interest, whether at the individual or the corporate level, always works out to everyone's advantage. Morality, in our view, requires a willingness to place some restraints on the pursuit of private interests by both individuals and political, economic, and professional groups. Accepting some such constraints is presupposed in what is meant by moral concern. Engineering ethics has as one task uncovering the moral limits of the pursuit of self-interest within the profession of engineering.

Religious Bases for Right Action

Our earlier selection of theories of right action did not include specific reference to moral imperatives based on religious belief. This was not an intentional slight, but rather a recognition of the fact that religious motivation can be the source of right action under any of the foregoing theories. Yet it need not be the sole source, and religious commitment is not synonymous with moral concern. It would surely be wrong to equate atheism, agnosticism, or a lack of interest in formal religious institutions with a lack of moral concern. It may be equally wrong to equate them with an absence of religious concern.

"Religion" can mean many things, from the quest for the metaphysical ultimates in the universe to the religious institutions which have grown around this quest. Religious practice can mean anything from passive participation in religious services to translation of a religious belief into right action. For our purposes the following wide definition, given by C. J. Ducasse, is useful:

> A religion . . . is any set of articles of faith—together with the observances, attitudes, obligations, and feelings tied up therewith—which, in so far as it is influential in a person, tends to perform two functions, one social and the other personal. The social function is to provide motivation for the individual to conduct himself altruistically on occasions when his individual interest conflicts with that of society and when neither his spontaneous altruistic impulses, nor the sanctions of the laws or of public opinion, are potent enough by themselves or together to motivate such conduct. The personal function, on the other hand, is to give the individual in some measure the serene assurance out of which flows courage on occasions of fear, endurance in adversity, strength in moments of weakness, dignity in defeat, humility in success, conscientiousness and moderation in the exercise of power (Ducasse, 415).

Perhaps the term "faith" implies more certitude than many religious seekers may have, in which case the term "hope" may not be amiss. In either case the words imply trust; trust that we can receive insight into what should govern right action and that we can be sustained in that action.

God-centered, or theistic, religions such as Judaism, Christianity, and Islam embrace a faith in a deity and derive therefrom more or less explicit divine laws. There are other religions, such as Buddhism, Taoism, and Confucianism, which do not involve a belief in a god but which nevertheless call for faith in a *right path* from which is derived a code of ethics. In Buddhism, for instance, the right path incorporates eight steps: right understanding, right intention, right speech, right action, right livelihood, right effort, right mindfulness, and right concentration. Part of the right path for many nontheistic religions is the need for contemplation and a feeling of oneness with nature in place of communion with a deity.

In any case, to sum up, as Ducasse points out the main social function of religion is to motivate right action, which involves the notion of ethics per se. Likewise the personal function of religion is very important. It has sustained many people trying to follow their convictions, and it can promote tolerance and moral concern for others when those motivated by it are confronted with the wide variety of beliefs and individual needs to be found in the world. Many engineers are certainly among those so motivated, which is why these paragraphs on religion are an appropriate part of our larger topic of engineering ethics.

Testing Ethical Theories

Our intent has not been to evaluate which of the major ethical theories is best. But we should mention several ways an ethical theory may be tested for adequacy. There are at least five widely used tests.

First, the theory must be clear and formulated with concepts which are coherent and applicable. Second, it must be internally consistent in that none of its tenets contradict each other. Third, neither the theory nor its defense can rely upon false information. Fourth, it must be sufficiently comprehensive to provide guidance in specific situations of interest to us.

Fifth, and perhaps most important, it must be compatible with our most carefully considered moral convictions about concrete situations and cases. If an abstract ethical theory said it was all right to torture mentally handicapped children to make other people happy, that would be fully sufficient to show the theory was false. Good theories, of course, may lead us upon reflection to modify some of our previously held views—one of their main uses is to correct mistaken views. In this way theories and concrete intuitions and judgments mutually interact. Ethical theories are developed to illuminate, unify, and correct our prereflective judgments; and moral experience and judgments about specific situations and cases are used to test ethical theories (Rawls, 46–53).

Study Questions

1 Find an article in a current newspaper or magazine which raises what you view as a moral issue in engineering. State the issue in your own words, making clear why you view it as a moral one. Explain how the problem might be approached and perhaps resolved by drawing on one of the goal-, duty-, or rights-based theories sketched in this section.
2 Apply each of the three main types of theories in responding to the following imaginary cases. In applying utilitarianism be sure to consider both act-utilitarianism and rule-utilitarianism. Then present your own answer—that is, with which theory do you agree?
   a A train is approaching a switch, and it is traveling too fast to stop before a tragedy occurs. Tied to one fork of the track are the leaders of three important nations (which you consider as promoting world peace and prosperity). Tied to the other fork are four people who are your closest friends and relatives, but who have no international or even national social importance. If you were in control of the switch, which fork in the track ought you to select?

b A doctor can save the lives of three important national leaders by making transplants of the kidneys and heart of a local convicted mass-murderer who is serving a life sentence. The operations would be done in secret and would involve the full cooperation of the local police officials, who would claim the murderer has been killed while trying to escape from prison. Should the operations be permitted?

3 Select one or more of the examples given in the first section of this chapter, "Moral Dilemmas." Drawing on one of the major ethical theories, explain how you would suggest the dilemma should have been resolved.

4 Consult the writings of a major ethicist and summarize the main ideas of the theories involved. [For suggested sources see the Bibliographical entries on Aristotle, Brandt, Kant, Melden, Mill, Oldenquist (1979), Rawls, and W. D. Ross. For helpful secondary sources, see Frankena (1973), Raphael, and Taylor.]

RELATIVISM AND SKEPTICISM

Can one really give good objective reasons for deciding what morally ought to be done within engineering practice—or anywhere else for that matter? Do not the ethical theories presented in the last section amount to little more than emotional prejudices? Perhaps moral values are ultimately arbitrary, which might explain why moral disputes seem unending. Surely we can all agree that people's moral outlooks often differ according to subjective factors like upbringing, cultural background, particular interests, and personal priorities.

Moral values, it might be continued, are thus *relative* to both conventional standards and individual preferences. Since engineering within large bureaucracies leaves little room for individual preferences, engineering ethics is said by some to be entirely relative to those conventions which dominate at any given time and place. Ethics in engineering should then be reducible to the codes and professional guidelines that set the conventions, which conventions in turn may represent little more than the whims of the time.

This appeal to relativism as it stands amounts to a radical ethical skepticism, and it will not do. Each of us has probably experienced skeptical doubts about the rationality of moral values. Those doubts are entirely natural in a century where despair and nihilism have had something of a heyday. Yet doubts themselves can be reasonable or unreasonable. And unless we simply stop reasoning altogether, even a complete skepticism requires a cogent defense.

There are many versions of relativism, depending upon how and to what moral values are supposed to be relative (Taylor, 13–30). We shall examine four of them, then conclude the section with two examples of actual situations showing how relativity of values can affect engineering practice. Some versions, such as the first two we shall discuss, are undeniably true and are obviously consistent with the possibility of an objective ethics based on moral concern. Others, however, such as the latter two we shall consider, are either of dubious truth or their implications must be carefully considered, for they appear to lend scepticism a firmer foundation than it may in fact have.

Descriptive Relativism

One type of relativism consists in factual or descriptive claims about people's moral outlooks, and hence is called *descriptive relativism*. In its simplest version it asserts two things: (1) There is widespread and sometimes fundamental disagreement about moral issues, and (2) the formation of our moral views is greatly influenced by our parents, peers, and culture.

These claims are obviously true. There have been, to be sure, serious debates among anthropologists about just how fundamental the differences in moral outlook are among various cultures. On the surface those differences seem enormous, especially when we focus on practices like head-hunting, human sacrifice, and cannibalism. Yet virtually all cultures have shown a commitment to protecting their members against needless death and suffering and to promoting social cooperation. And beneath surface differences in practice often lie disagreements over interpretation of facts or differences in circumstances rather than basic moral differences.

For example, the Eskimos and some American Indians used to leave their elderly to die. This seems radically inconsistent with our notions of respect for human life. Yet the Eskimos and Indians lived under harsh conditions requiring migration and rationing if the group was to survive. In such circumstances the elderly might voluntarily agree to abandonment out of respect for others and a wish not to become burdens. Or again, the Incas' practice of human sacrifice seems at first glance a sign of cruelty and lack of concern for life. But a deeper examination reveals that they believed their gods required such sacrifice to ensure the survival of their people as a whole, and it was considered an honor for the victims.

The important point is that mere diversity in what people believe hardly shows there is no objective truth. The fact that nearly everyone once believed the earth to be flat is irrelevant to determining the earth's actual shape. Likewise, the mere fact that other cultures accept moral beliefs differing from our own does not mean there is no objective moral truth and does not necessarily lead to moral scepticism.

Relationalism

There is a second relativistic claim which most of us will readily endorse. The claim is simply that moral judgments are properly made *in relation to* a wide variety of variable factors. This makes it usually impossible to formulate absolute, exceptionless, specific rules like "lying is always and everywhere immoral." There can arise special circumstances where people might have to lie to save a life or to protect their privacy from mischievous intruders. If we need a label for this view, we might call it *relationalism*.

Virtually all ethical theorists, with the exception of Kant and some theistic ethicists, accept relationalism as true. Ross, for example, created his concept of prima facie duties with this in mind. He sought to emphasize how everyday rules such as "Don't lie" and "Keep promises" can have exceptions when they come into conflict with other rules and cause moral dilemmas. Utilitarians emphasize that the context in which an act is performed (act-utilitarianism) or in which a rule operates (rule-utilitarianism) must be examined in assessing good and bad consequences. And rights-based theories allow for the possibility of many morally relevant considerations in determining when it is appropriate to exercise a moral right.

Relationalism, we should note, allows for the fact that the special customs of other cultures are sometimes morally relevant considerations. In Japan, for example, to show respect for a traditional Japanese family one should embrace the custom of removing shoes before entering the house. Relationalism needs to be borne in mind in regard to multinational engineering contexts involving the very different conventions to be encountered around the world, as we shall see in a moment.

Relationalism only says that customs are morally *relevant* considerations which need to be taken into account. It does not say that foreign customs always dictate the right thing to do. This latter, more extreme view, leads to the next type of relativistic claim we shall consider.

Ethical Conventionalism

The preceding versions of relativism must be sharply distinguished from a disturbing and dubious version called *ethical conventionalism*. It asserts that whatever conventions a group adopts are thereby automatically right. Customs, laws, and other conventions are claimed to be decisive in determining what ought to be done. To use an extreme illustration, this view would allow that Hitler and his followers acted rightly when they murdered six million Jews, for they believed they were doing something right and their customs involved fanatical antisemitism.

Conventionalism makes objective moral reasoning and the use of ethical theories unnecessary. By the same token it also makes nonsense of our common understanding of morality, which allows for the fact that we can and in some cases ought to raise moral objections to current practices and conventions.

Defenders of conventionalism generally add that a belief or action was right "for them," though not "for us." But this is a needless shuffle. If the expression "right for them" means merely that they believed something was right and that this belief played a key role in their lives (such as when we say of the ancients that the earth was "flat for them"), then this shaky view reduces to the truism that they believed what they believed. If, by contrast, it means their beliefs justified their actions, then we are left with the false claim that believing something to be right makes it right. Beliefs, however customary or widely shared, are not self-certifying. Nor, we might add, are attitudes concerning morality.

Conceptual Relativism

Conceptual relativism is the view that truth is relative to conceptual schemas. A conceptual schema is a set of fundamental concepts, assumptions, and procedures for justifying views. As applied to morality, conceptual relativism says that the meaning, truth, and justification of moral judgments must be determined within a given fundamental moral perspective (e.g., a rights-based perspective).

This view, then, holds that it is at least possible that there be different moral perspectives which are in some sense equally valid although they may be incompatible with each other. Hence what is true within one moral framework (e.g., "There are human rights") might not be true within another, and there may be no way to determine which framework is more valid. In addition, according to conceptual relativism, moral concepts and assumptions develop within "ways of life" involving specific patterns of attitudes, emotional responses, and behavioral interactions among people. Thus moral truth and justification may ultimately vary with differences in those ways of life, and it may not be possible to determine which basic life pattern is the right or best one.

We shall not attempt to assess this subtle version of relativism, a version very much under debate within philosophy and other disciplines. Instead we shall rest content with three observations about it. First, conceptual relativism is not solely a theory about morality. It applies also to fundamentally differing conceptual schemas in mathematics (such as Euclidian versus non-Euclidian geometries), logic, and science. Hence it should not be interpreted as casting ethics into any more dubious or less respectable status than these other disciplines.

Second, conceptual relativism allows for a significant type of objectivity within general conceptual schemas. For example, once moral rights are included in one's conceptual schema it becomes objectively right or wrong to do certain things. Objective judgments about right and wrong can then be made against this background of rights.

Third, in ethics as elsewhere it pays not to be too hasty in concluding that irresolvable differences exist among major conceptual schemas. Surface differences often turn out to be more a matter of different emphasis than of radical incompatibility. Most of the ethical theories examined in Chap. 1, as we saw, agree in giving some role to duties, moral rights, and good consequences. Thus these concepts are widely shared components of very different moral theories and conceptual schemas. Where genuine differences do exist, further reflection sometimes reveals how to establish one theory as better than another (using the criteria given in the previous section of this chapter for testing theories).

Multinational Engineering

Questions about the relativity of values arise in connection with problems faced by engineers and engineering firms practicing in foreign countries. Suppose, as is often the case, that the foreign country's customs or engineering standards bearing on morality are significantly different from those of one's native country. What standards should guide one's conduct?

Ethical conventionalism might support the maxim, "When in Rome do as the Romans do." That is, it would have us believe there is no real problem: One should just drift with the conventions dominant in the local area. An equally extreme opposite view would have one retain precisely the same specific practices endorsed at home, never making any adjustments to the new culture. Both of these choices, however, seem to us unacceptable.

Consider first the system of racial separation, or apartheid, in the Republic of South Africa (Callaghan, 325–359). The economic aspects of the system include differential wage scales, with black workers paid considerably less than the wages paid to whites for similar work. Advancements to high managerial positions within companies have either been limited to whites or made much more difficult for black employees. Segregation has extended to rest rooms, lounge areas, and assembly lines.

A number of international engineering firms do business in South Africa. Most of them, although not all, have in the past adopted at least some of the South African policies. Was this morally permissible? Were they perhaps even obligated to conform to the dominant foreign customs?

We take the view that it was neither obligatory nor, in most cases, permissible to do so. Racial discrimination was wrong in the United States even during the time it was accepted by the majority of citizens. It is equally wrong now, and seriously so. Few practices so clearly violate fundamental moral principles about human equality, dignity, and rights.

The major ethical theories can be invoked to support this judgment. From Kant's duty-based perspective, the practice constitutes treating the black Africans as a mere means to promoting the advantages of the white minority and foreign corporations seeking inexpensive labor. From Locke's rights-based perspective, it amounts to a blatant infringement of human rights. And from a utilitarian perspective, the immense suffering of the black majority in South Africa outweighs the financial benefits of the already advantaged groups, so that the practice does not produce the most good for the most people.

Is our harsh judgment applicable in all instances? Might there be mitigating circumstances? Our earlier comments on relationalism would require us to look more closely at the practices of particular American companies before making unqualified condemnations. Are they taking steps to increase wages for black workers following a definite timetable so that equal work will eventually be rewarded by equal pay? Are they working toward bringing more blacks into management positions by offering them the required training now? To the extent that such efforts are underway and indeed constitute more than tokenism, any judgment of blame would have to be qualified.

Now let us consider a different sort of case. Engineering firms and other companies doing business in Italy are confronted with a tax system which strikingly departs from the American one. According to one account, corporations routinely submit a tax return showing profits of only 30 to 70 percent of their actual levels (Kelly, 37–39). Government officials generally estimate any corporation's taxes at several times over what that corporation claims it owes. Then the two sides negotiate the final figure as a compromise. The Italian revenue agents take as their "fee" an unreported cash payment directly from the representative for the corporation. The size of this payment influences how much or how little the revenue agent will demand in taxes for the Italian government.

Anything like such a practice, of course, would be considered corrupt in the United States. It involves misrepresenting actual profits, secretive payments, and an open door to biased tax assessments. Some would describe it in terms of deception, dishonesty, bribes, and conflicts of interest. Yet companies doing business in Italy which accurately report their profits would be assessed taxes much higher than they deserve. It would simply be unworkable to try doing business while refusing to participate in the practice.

Of course such "pragmatic" considerations concerning workability do not automatically justify participation in a practice. But several features should be noted that set this case apart from the South African example.

First, the practice is both officially sanctioned by the government and also tolerated by the Italian people. A shared understanding exists as to its legitimacy within Italy. Second, in some respects the practice is analogous to bargaining in the market square and bluffing in poker. It could at least be argued that sometimes deception is permissible when all participants accept it and no serious moral harm comes from it. Third, and perhaps most important, no basic human rights are violated by the practice. To be sure, the custom is in many ways undesirable because it lends itself to abuses our own system tries to discourage. And doubtless it encourages laxity about truthfulness in public life and may lessen the public trust needed for effective government. Yet it is not obvious that foreign companies participating in it are actually doing any harm.

These two examples suggest that it is neither always right nor always wrong to work within the systems of other countries when those systems from our point of view involve undesirable customs. In general, foreign customs provide morally relevant data which must enter into our ethical deliberations. But neither they nor current U.S. practice can be assumed to dictate automatically how engineers and corporations ought to act within unusual social settings. Each case, or at least each type of case, must be examined on its own merits.

Study Questions

1 During 1972 and 1973 the President of Lockheed, A. Carl Kotchian, authorized secret payments totaling around $12 million beyond a contract to representatives of Japan's Prime Minister Tanaka. Later revelations of the bribes helped lead to the resignation of Tanaka and also to new laws in this country forbidding such payments. Mr. Kotchian believed at that time it was the only way to assure sales of Lockheed's TriStar airplanes in a much needed market.

In explaining his actions, Mr. Kotchian cited the following facts (Kotchian, 67–75): (1) There was no doubt in his mind that the only way to make the sales was to make the payments. (2) No U.S. law at the time forbade the payments. (3) The payments were financially worthwhile, for they totalled only 3 percent of an expected $430 million income for Lockheed. (4) The sales would prevent Lockheed layoffs, provide new jobs, and thereby benefit workers' families and their communities as well as the stockholders. (5) He himself did not personally initiate any of the payments, which were all requested by Japanese negotiators. (6) In order to give the TriStar a chance to prove itself in Japan, he felt he had to "follow the functioning system" of Japan. That is, he viewed the secretive payments as the accepted practice in Japan's government circles for this type of sale.

   a Drawing on the distinctions made in this section, explain the several senses in which someone might claim that how Mr. Kotchian ought to have acted is a "relative" matter. Which of these senses, in your view, would yield true claims and which false?

   b Develop the strongest moral argument you can think of in favor of Mr. Kotchian's actions, then the strongest argument you can against his actions. Draw upon the ethical theories presented in the previous section: either act-utilitarianism, rule-utilitarianism, Kant or Ross's duty-based theory, or Locke's rights-based theory. Then indicate which argument you find most compelling and why you disagree with the opposing view.

2 Some American companies have refused to promote women into positions of high authority in their international operations in Asia, the Middle East, and South America. Their rationale is that business will be hurt because some foreign customers do not wish to deal with women. It might be contended that this practice is justified out of respect for the customs of countries which discourage women from entering business and the professions.

Circuit Judge Warren J. Ferguson argued, however, that such practices are wrong. He ruled that sex stereotypes are not to be used in formulating job qualifications, and that customer preferences do not justify sex discrimination. He added that while our legal system cannot be used to force other countries to stop sex discrimination, other countries cannot dictate sex discrimination for citizens of our country (Mayer, 2).

Present and defend your view as to whether Judge Ferguson's ruling is morally justified.

MOTIVES AND RESPONSIBILITY

We have been focusing on judgments about right and wrong actions. Other types of evaluative judgments, however, are directed toward people. We sometimes view people as good or bad, virtuous or vicious, responsible or irresponsible, and so on. In order to avoid confusion, it is important to identify at least some of the differences and connections between judgments about people and judgments about actions.

Motives

Many moral evaluations of people center upon their motives and intentions. Saying that individuals are good or have good characters, for example, typically means they are motivated to avoid wrong-doing and to be considerate of others. Likewise, calling an action good can mean it is not only right but performed out of good motives.

What makes a motive good, and a person deserving of esteem, respect, or admiration? Kant's answer was that only actions done out of a sense of duty warrant moral esteem (Kant, 16–19). As an illustration he described a businessman who is honest merely because honesty helps to promote his business interests. The man follows the maxims "Honesty is the best policy" and "Honesty pays" in not overcharging children, knowing that their parents might find out if he cheated them and this might hurt his business. The man's action, Kant said, "conforms with duty"—we might say he did the dutiful thing. But he deserves no special moral respect or esteem since he was motivated by prudence rather than an active concern to do what is dutiful.

Kant's comment on this example is accurate, but his general view is too narrow. We hold people in high esteem when they act out of motives other than a stern respect for duty. It seems perfectly appropriate to respect people for acting out of spontaneous love, friendship, and benevolence. In fact, sometimes we value such acts because they go beyond what duty requires, especially when they attain heroic or saintly proportions. Later we will ask whether some whistleblowers, who at great expense to their own careers informed the public of dangers, might not fall into this category.

In contrast with situations where the right thing is done from nonmoral motives, actions may be wrong even though performed out of morally good motives. Suppose that out of friendship a self-employed engineer who is leaving on vacation loans his business car to a trusted colleague. Unknown to the engineer, the friend is seriously off-balance because he recently broke up with his girl friend. While he has the car he gets drunk, drives through a highway barrier, and kills himself and several others. It was wrong to loan him the car. But his good intentions and motives, together with the limited knowledge available to him about his friend's condition, excuse the lender from blame.

This example makes clear how judging a person's actions as wrong is not the same as blaming the person. To blame is to condemn a person rather than to appraise the moral status of an action. Blaming is itself an action subject to moral appraisal as to its appropriateness or inappropriateness .

Motives and Professional Ethics   How important are motives within professional ethics? Does it make sense to say that people can be ethical professionals even though they do not show much moral concern for others?

This can be possible, at least to some degree. In his novel *A Burnt-Out Case*, Graham Greene describes an architect who has reached the top of his profession without caring very much about the good of the public which has benefited from his work. The architect, world renowned, abandons a career in which he has made numerous brilliant contributions without any wrong-doing. He travels to Africa and meets a doctor who is practicing medicine on the basis of a concern for his patients. In one scene the architect explains to the doctor that his interest had always been in the "space and the light and the proportion" of buildings, not in the people who might use them (Green, 44). Jokingly, the doctor remarks that he would not have trusted the plumbing in the structure designed by the architect. But the architect presses his point. He confesses that of course he had to consider human needs, but only in the same way he had to consider the brick, glass, and other building materials. His sole motivation, however, was the creation of beautiful structures.

As this incident suggests, it is possible for a person to act as a responsible architect or engineer from primarily nonmoral motives, such as a sheer pleasure in the beauty and an excitement over the technical aspects of the work. The doctor's facetious remark about plumbing reminds us that architects and engineers must be concerned professionally to satisfy the needs of those they serve, but that concern need not necessarily be moral in origin. Accordingly, much of professional ethics focuses on the level of care the architect says he showed in his work—respect for the safety, well-being, and needs of those affected by the professional activities involved—no matter what the ultimate wellspring of that care may be.

Nevertheless, the question of ultimate motivation will enter from time to time into discussions of professional ethics. It is virtually inevitable that professionals will form general appraisals of the characters and basic concerns of themselves and their colleagues. And on occasion they may even be called upon as members of professional societies to pass judgments on colleagues charged with misconduct. Such judgments typically involve some consideration of motives. Certainly an individual's own self-identity will involve some understanding of his or her ultimate personal commitments and motives for professional activities. We will return to this matter in Chap. 7 while discussing career choices.

Moreover, while nonmoral motives often lead to morally right actions, they do not always do so. They may also lead to "cutting corners" where moral considerations are involved and eventually to a major neglect of professional duties.

Moral Responsibility

The notion of moral responsibility cuts across judgments about both right actions and people. In every case where moral responsibility is ascribed to someone, a moral judgment is being made; judgments may be of various types (Hart, 211–230). The interest may be in assessing (1) obligations to perform right actions, (2) general moral capacities of people, (3) the virtue of a person, or (4) liabilities and accountability for actions.

1 We speak of persons as *having* moral responsibilities. In this sense, responsibilities are simply obligations and duties to perform morally right acts. Some of those are shared by us all: for example, the responsibilities to be truthful, to be fair, and to promote justice. Others relate only to people within certain social roles or professions: For example parents have specific responsibilities to care for their children, a safety engineer might have responsibilities for making regular inspections at a building site, or an operations engineer might have special responsibilities for identifying potential benefits and risks of one system as compared to another.

2 Sometimes when we ascribe responsibility to a person viewed as a whole rather than in respect to a specific area of his or her conduct, we have in mind an active *capacity* for knowing how to act in morally appropriate ways. In this sense young children are not yet morally responsible. They gradually become so as they mature and learn how to be responsive to the needs and interests of others. Adult sociopaths who lack any sense of guilt for wrong-doing never become responsible in this sense.

3 At other times when we say someone is responsible, we mean to ascribe a general moral *virtue* to the person. We mean that he or she is regularly concerned to do the right thing, is conscientious and diligent in meeting obligations, and is someone who can be counted on to carry out duties or be considerate of others. In Chap. 3 we will develop a fuller conception of moral responsibility, viewed as a general virtue, on the part of engineers.

4 Finally, "responsible" often means accountable, answerable, or liable for meeting obligations. In this sense, to say individuals are responsible for actions means they can be "held to account" for them: that is, they can be called upon to explain why they acted as they did; to provide excuses or justification, if appropriate; and to be open to commendation or censure, praise or blame, or demands for compensation. We also hold ourselves accountable for our own actions, responding to them with emotions of self-esteem or shame, self-respect or guilt. This notion of responsibility will also be developed more fully in Chap. 3.

Accountability and Voluntary Action   Usually when we hold a person accountable for an action we imply that the action was not completely involuntary. But it is not easy to know precisely what this requirement amounts to.

Aristotle suggested that involuntary acts are of two main kinds (Aristotle, 964–967). First, they include acts done in ignorance. Because of ignorance, the person mentioned earlier who loaned his car to a friend did not voluntarily contribute to the death of the friend. Yet the problem is that we also hold people accountable when they *should have known* what they were doing and what the likely consequences of their action would be. Ignorance then, is not always an excuse. On the other hand, it is often difficult to judge fairly what people should have known, especially within the types of complicated situations that can arise in professions like engineering.

Second, Aristotle said acts are involuntary when performed under compulsion. He interpreted compulsion as an external force which determines our actions. Aristotle also noted that the mere existence of obstacles does not entirely negate voluntariness. Rather their presence limits the range of choices open to us. When those choices become sufficiently limited, we have to think in terms of degrees of voluntariness, and hence degrees of liability for harm done. An example would be the limitations placed on our decisions in response to a kidnapper who demands a ransom for not killing someone dear to us.

To make things even more complicated, many psychologists today would make a major addition to Aristotle's views: Acts may be involuntary when they are generated by uncontrollable inner compulsions, such as those motivating psychotics or pathological liars. Yet it is extremely difficult to make accurate assessments of other people at this level.

Causal and Legal Responsibility There are two other concepts of responsibility (Hart, 214–215). These should not be confused with moral responsibility in any of its four preceding senses. First, *causal responsibility* consists simply in being a cause of some event. In this sense we speak of lightning as being responsible for a house catching fire.

People can be causally responsible for an event without necessarily being morally responsible for it. For example, a 2-year-old child may cause a fire while playing with matches, but it is the parents who left the matches within the child's reach who are morally responsible for the fire.

Second, *legal responsibility* should also be distinguished from moral responsibility. An engineer or engineering firm can be held legally responsible for harm which was so unlikely and unforeseeable that little or no moral responsibility is involved.

One famous court case involved a farmer who lost an eye when a metal chip flew off the hammer he was using (Vaughn, 41–47). He had used the hammer without problems for 11 months before the accident. It was constructed from metals satisfying all the relevant safety regulations, and no specific defect was found in it. The manufacturer was held legally responsible and required to pay damages. The basis for the ruling was the doctrine of "strict legal liability," which does not require proof of defect or negligence in design. Yet surely the manufacturer was not morally culpable or blameworthy for the harm done. If we say the manufacturer was morally responsible, we mean at most that the company has an obligation (based on the special relationship between it and the farmer created by the accident) to help remedy the problem caused by the defective hammer.

Conversely, it is also possible to be morally responsible for something one cannot be held legally responsible for. For example, because of the fine wording of a contract an engineer may be free from any legal liability for failing to report an observed danger at a construction site. Yet it may have been his or her professional and moral obligation to report that danger.

Study Questions

Kermit Vandivier had worked at B. F. Goodrich for 5 years, first in instrumentation and later as a data analyst and technical writer. In 1968 he was assigned to write a report on the performance of the Goodrich wheels and brakes commissioned by the Air Force for its new A7D light attack aircraft. According to his account, he became aware of the design's limitations and of serious irregularities in the qualification tests. The brake failed to meet Air Force specifications. Upon pointing out these problems, however, he was given a direct order to stop complaining and write a report which would show the brake qualified. He was led to believe that several layers of management were behind this demand and would accept whatever distortions might be needed because their engineering judgment assured them the brake was acceptable.

Vandivier then drafted a 200-page report with dozens of falsifications and misrepresentations. But he refused to sign it. Later he gave as excuses for his complicity the facts that he was 42 years old with a wife and six children. He had recently bought a home and felt financially unable to change jobs. He felt certain that he would have been fired if he had refused to participate in writing the report (Vandivier, 20–24).

1 Present and defend your view as to whether Vandivier was justified in writing the report or not. In doing so, draw upon one of the theories of right action discussed in the second section of this chapter.
2 Was Vandivier guilty or blameworthy? That is, even if his actions were wrong, is it appropriate to excuse him from blame because of circumstances beyond his control?
3 Is Vandivier responsible for what he did? In answering this question, distinguish between the various senses of "responsible" discussed in this section.

SUMMARY

Moral problems, in the widest sense, are any situations calling for decisions based upon moral considerations. Sometimes what ought to be done is a straightforward matter, and the only difficulty is in avoiding temptations to violate moral obligations. At other times it may be unclear whether a moral principle applies, as in deciding whether accepting some gifts from salespeople violates the rule "Do not accept bribes." *Moral dilemmas* are those moral problems in which two or more moral obligations, duties, rights, ideals, or applications of a single principle come into conflict in a given situation and it seems that not all of them can be respected or fulfilled.

Philosophers have suggested a variety of theories about how to characterize what makes actions right or wrong. Some of the more important theories of right action can be summarized as follows:

Acts are morally right when:

• They produce the most good for the most people.	Act-utilitarianism	Goal-based theories (good consequences are basic)
• They fall under a rule which if widely followed would produce the most good for the most people.	Rule-utilitarianism	

• They fall under rules which can be willed, without inconsistency, for everyone to follow.	Kant's universalizing test	
• They fall under only one (prima facie) duty (e.g., to be fair) or where two duties conflict they fall under the duty which reflection reveals as most important in that situation.	Ross	Duty-based theories (duties are basic)
• They fall under principles which would be agreed upon by all rational agents in a hypothetical contracting situation.	Rawls	
• They respect others' moral rights or involve exercising one's own rights in morally appropriate ways.	Locke	Rights-based theories (rights are basic)

There is a question about whether *rational egoism,* which says people ought to pursue only their own individual good, should be viewed as an ethical theory. For morality seems to involve a concern for more than just one's own private self-interest. Religious commitments should not be viewed as competing with the main moral theories, but rather as providing one source of motivation for being ethical.

Ethical skepticism becomes less tempting by distinguishing between several senses in which morality is sometimes claimed to be "relative":

• *Descriptive relativism:* Factually speaking, there is a wide diversity in beliefs concerning morality, and these differences are culturally influenced and sometimes of a fundamental nature.

• *Relationalism:* Moral judgments are made in relation to special factors which may differ from case to case in a way that usually makes it impossible to formulate exceptionless, universal, simple rules like "Never lie."

• *Ethical conventionalism:* Moral rightness consists in following the dominant conventions or customs of a given region, culture, or era and wrongness lies in departing from them.

• *Conceptual relativism:* Truth, including moral truth, is dependent upon and relative to specific conceptual schemas, which are sets of basic concepts, assumptions, and ways of justifying intellectual or moral claims.

Of all these views, only ethical conventionalism clearly undermines the possibility of objective moral evaluations. Relationalism identifies an important aspect that affects making moral judgments; it also allows for the fact that foreign customs are germane, although not always decisive, factors in the ethical conduct of engineers working within a multinational context.

Finally, if *actions* can be judged right or wrong, *people* can be judged as good or bad, virtuous or vicious, responsible or irresponsible. Such evaluations of people involve an emphasis on motives and intentions. And judgments of moral responsibility must be carefully distinguished from descriptions of causal responsibility or legal responsibility. The concept of moral responsibility can be applied in different ways to refer to (1) obligations ("responsibilities" in the strict sense), (2) the general capacity to act in morally concerned ways, (3) the virtue of being conscientiously concerned to meet one's moral obligations, (4) accountability for actions.

PART II

THE EXPERIMENTAL NATURE OF ENGINEERING

To undertake a great work, and especially a work of a novel type, means carrying out an experiment. It means taking up a struggle with the forces of nature without the assurance of emerging as the victor after the first attack.

Louis Marie Henri Navier (1785–1836)
a founder of structural analysis

A ship in harbor is safe, but that is not what ships are built for.

John A. Shedd

CHAPTER 3

ENGINEERING AS SOCIAL EXPERIMENTATION

As it departed on its first voyage in April 1912, the *Titanic* was proclaimed the greatest engineering achievement ever. Not merely was it the largest ship the world had seen, having a length of eight football fields, and the most glamorous of ocean liners, complete with the first seagoing masseuse. It was supposed to be the first fully safe ship. Since the worst collision envisaged was at the juncture of two of its sixteen watertight compartments, and since it could float with any four compartments flooded, the *Titanic* was confidently believed to be virtually unsinkable.

Buoyed by such confidence, the captain allowed the ship to sail full speed at night in an area of reported icebergs, one of which tore a 300-foot gash in its side, flooding five compartments. Time remained to evacuate the ship, but there were only enough lifeboats to accommodate half the passengers and crew. Then-current regulations regarding lifeboats did not take into account vessels of this size, and no extra precautions had seemed necessary for an unsinkable ship. The result was over 1500 people dead (Lord; Wade).

In his poem written shortly after the event, "The Convergence of the Twain," Thomas Hardy portrayed the meeting of the ship and iceberg as determined by unpredictable fate: "No mortal eye could see/The intimate welding of their later history." Yet greater imagination and prudence could have prevented the disaster.

Interestingly enough, novelists did not lack the imaginative foresight to describe scenarios which paralleled the later real events in shocking detail.

In Morgan Robertson's 1898 novel *Futility,* a ship almost identical in size to the *Titanic* was wrecked by an iceberg on a cold April night. The ship in the book was named the *Titan;* it too had a fewer than sufficient number of lifeboats. Mayn Clew Garnett's story "The White Ghost of Disaster" was being readied for publication in *Popular Magazine* while the *Titanic* was on her maiden voyage. It is said that Garnett had dreamed the story while traveling on the *Titanic's* sister ship, the *Olympic.* Again, circumstances similar to those surrounding the sinking of the *Titanic,* as well as an insufficient number of lifeboats to save all the passengers, were key elements in the narrative (Wade, 70–71).

The *Titanic* remains a haunting image of technological complacency. Perhaps all we can take for granted today is Murphy's law that if anything can go wrong, it will—sooner or later. All products of technology have some potential dangers, and thus engineering is an inherently risky activity. In order to underscore this fact and help in exploring its ethical implications, we suggest that engineering should be viewed as an experimental process. It is not, of course, an experiment conducted solely in a laboratory under controlled conditions. Rather, it is an experiment on a social scale involving human subjects.

ENGINEERING AS EXPERIMENTATION

Experimentation is commonly recognized to play an essential role in the design process. Preliminary tests or simulations are conducted from the time it is decided to convert a new engineering concept into its first rough design. Materials and processes are tried out, usually employing formal experimental techniques. Such tests serve as the basis for more detailed designs, which in turn are tested. At the production stage further tests are run, until a finished product evolves. The normal design process is thus iterative, carried out on trial designs with modifications being made on the basis of feedback information acquired from tests. Beyond those specific tests and experiments, however, each engineering project taken as a totality may itself be viewed as an experiment.

Similarities with Standard Experiments

Several features of virtually every kind of engineering practice combine to make it appropriate to view engineering projects as experiments. First, any project is carried out in partial ignorance. There are uncertainties in the abstract model used for the design calculations; there are uncertainties in the precise characteristics of the materials purchased; there are uncertainties about the nature of the stresses the finished product will encounter. Engineers do not have the luxury of waiting until all the relevant facts are in before commencing work. At some point theoretical exploration and laboratory testing must be bypassed for the sake of moving ahead on a project. Indeed, one talent crucial to an engineer's success lies precisely in the ability to accomplish tasks with only a partial knowledge of scientific laws about nature and society.

Second, the final outcomes of engineering projects, like those of experiments, are generally uncertain. Often in engineering it is not even known what the possible outcomes may be, and great risks may attend even seemingly benign projects. A reservoir may do damage to a region's social fabric or to its ecosystem. It may not even serve its intended purpose if the dam leaks or breaks. An aqueduct may bring about a population explosion in a region where it is the only source of water, creating dependency and vulnerability without adequate safeguards. An aircraft may become a status symbol which ultimately bankrupts its owners. A special-purpose finger-print reader may find its main application in the identification and surveil-lance of dissidents by totalitarian regimes. A nuclear reactor, the scaled-up version of a successful smaller model, may exhibit unexpected problems which endanger the surrounding population, leading to its untimely shut-down at great cost to owner and consumers alike. A hair dryer may expose the unknowing or unwary user to asbestos poisoning from the insulation in its barrel.

Third, effective engineering relies upon knowledge gained about prod-ucts both before and after they leave the factory—knowledge needed for improving current products and creating better ones. That is, on-going success in engineering depends upon gaining new knowledge, just as does on-going success in experimentation. Monitoring is thus as essential to engineering as it is to experimentation in general. To monitor is to make periodic observations and tests in order to check for both successful performance and unintended side effects. But since the ultimate test of a product's efficiency, safety, cost-effectiveness, environmental impact, and aesthetic value lies in how well that product functions within society, monitoring cannot be restricted to the development or testing phases of an engineering venture alone. It also extends to the stage of client use as well. Just as in experimentation, both the intermediate and final results of an engineering project deserve analysis if the correct lessons are to be learned from it.

Learning from the Past

It might be expected that engineers would learn not only from their own earlier design and operating results, but also from those of other engineers. Unfortunately that is frequently not the case. Lack of established channels of communication, misplaced pride in not asking for information, embarrassment at failure, and plain neglect often impede the flow of such information and lead to many repetitions of past mistakes. Here are a few examples:

1 The *Titanic* lacked a sufficient number of lifeboats decades after most of the passengers and crew on the steamship *Arctic* had perished because of the same problem (Wade, 417).

2 "Complete lack of protection against impact by shipping caused Sweden's worst ever bridge collapse on Friday as a result of which eight people were killed." Thus reported the *New Civil Engineer* on January 24, 1980. On May 15 of the same year it also reported the following: "Last Friday's disaster at Tampa Bay, Florida, was the largest and most tragic of a growing number of errant ships colliding with bridges over navigable waterways." While collisions of ships with bridges do occur, the Florida bridge was not designed with horizontal impact forces in mind because the code did not require it.

3 In June 1966 a section of the Milford Haven bridge in Wales collapsed during construction. A bridge of similar design was being erected by the same bridge builder (Freeman Fox and Partners) in Melbourne, Australia, when it too partially collapsed, killing 33 and injuring 19. This happened in October of the same year, shortly after chief construction engineer Jack Hindshaw (also a casualty) had assured worried workers that the bridge was safe (Yarrow Bridge, 415).

4 Valves are notorious for being among the least reliable components of hydraulic systems. It was a pressure relief valve, and lack of positive information regarding its open or shut state, which helped lead to the nuclear reactor accident at Three Mile Island on March 28, 1979. Similar malfunctions had occurred with identical valves on nuclear reactors at other locations. The required reports had been filed with Babcock and Wilcox, the reactor's manufacturer, but no attention had been given to them (Sugarman, 72).

5 The Bureau of Reclamation, which built the ill-fated Teton Dam, allowed it to be filled rapidly, thus failing to provide sufficient time to monitor for the presence of leaks in a project constructed out of less than ideal soil. The Bureau did not heed the lesson of its Fontenelle Dam, where 10 years earlier massive leaks had also developed and caused a partial collapse (Shaw; Boffey).

These examples, and others to be given in later chapters, illustrate why it is not sufficient for engineers to rely on handbooks alone. Engineering, just like experimentation, demands practitioners who remain alert and well informed at every stage of a project's history.

Contrasts with Standard Experiments

To be sure, engineering differs in some respects from standard experimentation. Some of those very differences help to highlight the engineer's special responsibilities. And exploring the differences can also aid our thinking about the moral responsibilities of all those engaged in engineering.

Experimental Control   One great difference is that of experimental control. In a standard experiment this involves the selection, at random, of members for two different groups. The members of one group receive the special, experimental treatment. Members of the other group, called the *control group,* do not receive that special treatment although they are subjected to the same environment as the first group in every other respect.

In engineering this is not the usual practice, unless the project is confined to laboratory experimentation, because the experimental subjects are humans out of the range of the experimenter's control. Indeed, clients and consumers exercise most of the control because it is they who choose the product or item they wish to use. This makes it impossible to obtain a random selection of participants from various groups. Nor can parallel control groups be established based on random sampling. Thus no careful study of the effects of changing variables on two or more comparison groups is possible, and one must simply work with the available historical and retrospective data about various groups which use the product.

This suggests that the view of engineering as a social experiment involves a somewhat extended usage of the concept of experimentation. Nevertheless, "engineering as social experimentation" should not be dismissed as a merely metaphorical notion. There are other fields where it is not uncommon to speak of experiments whose original purpose was not experimental in nature and which involve no control groups.

For example, social scientists monitor and collect data on differences and similarities among existing educational systems that were not initially set up as systematic experiments. In doing so they regard the current diversity of systems as constituting what has been called a "natural experiment" (as opposed to a deliberately initiated one) (Rivlin, 70). Similarly, we think that engineering can be appropriately viewed as just such a "natural experiment" using human subjects, despite the fact that most engineers do not currently consider it in that light.

Informed Consent   Viewing engineering as an experiment on a societal scale places the focus where it should be: on the human beings affected by technology. For the experiment is performed on persons, not on inanimate objects. In this respect, albeit on a much larger scale, engineering closely parallels medical testing of new drugs and techniques on human subjects.

Society has recently come to recognize the primacy of the subject's safety and freedom of choice about whether to participate in medical experiments. Ever since the revelations of prison and concentration camp horrors in the name of medicine, an increasing number of moral and legal safeguards have arisen to ensure that subjects in experiments participate on the basis of informed consent.

But while current medical practice has increasingly tended to accept as fundamental the subject's moral and legal rights to give informed consent before participating in an experiment, contemporary engineering practice is only beginning to recognize those rights. We believe that the problem of informed consent, which is so vital to the concept of a properly conducted experiment involving human subjects, should be the keystone in the interaction between engineers and the public. We are talking about the *lay public*. When a manufacturer sells a new device to a knowledgeable firm which has its own engineering staff, there is usually an agreement regarding the shared risks and benefits of trying out the technological innovation.

Informed consent is understood as including two main elements: knowledge and voluntariness. First, subjects should be given not only the information they request, but all the information which is needed for making a reasonable decision. Second, subjects must enter into the experiment without being subjected to force, fraud, or deception.

The mere purchase of a product does not constitute informed consent, any more than does the act of showing up on the occasion of a medical examination. The public and clients must be given information about the practical risks and benefits of the product in terms they can understand. Supplying complete information about the product is neither necessary nor in most cases possible. In both medicine and engineering there may be an enormous gap between the experimenter's and the subject's understanding of the complexities of an experiment. But while this gap most likely cannot be closed, it should be possible to convey all pertinent information needed for making a reasonable decision about whether to participate or not.

We do not propose a proliferation of lengthy environmental impact reports. We favor the kind of sound advice a responsible physician gives a patient when prescribing a course of drug treatment which has possible side effects. The physician must search beyond the typical sales brochures from drug manufacturers for adequate information; hospital management must allow the physician the freedom to undertake different treatments for different patients, as each case may constitute a different "experiment" involving different circumstances; finally, the patient must be readied to receive the information.

Likewise, an engineer cannot succeed in providing essential information about a project or product unless there is cooperation by management and a receptivity on the part of those who should have the information. Management is often understandably reluctant to provide more information than current laws require, fearing disclosure to potential competitors and exposure to potential lawsuits. Moreover, it is possible that, paralleling the experience in medicine, clients or the public may not be interested in all of the relevant information about an engineering project, at least not until a crisis looms. It is important nevertheless that all avenues for disseminating such information be kept open and ready.

We note that the matter of informed consent is surfacing indirectly in the current debate over acceptable forms of energy. Representatives of the nuclear industry can be heard expressing their impatience with critics who worry about reactor malfunction while engaging in statistically more hazardous activities such as driving automobiles and smoking cigarettes. But what is being overlooked by those representatives is the often-found human readiness to accept risks voluntarily undertaken as in daring sports, even while objecting to involuntary risks resulting from activities in which the individual is neither a direct participant nor a decision maker. In other words, we all prefer to be the subjects of our own experiments rather than those of somebody else. When it comes to approving a nearby oil drilling platform or a nuclear plant, affected parties expect their consent to be sought no less than it is when a doctor contemplates surgery.

Prior consultation of the kind suggested can be effective. When Northern States Power Company (Minnesota) was planning a new power plant, it got in touch with local citizens and environmental groups before it committed large sums of money to preliminary design studies. The company was able to present convincing evidence regarding the need for a new plant and then suggested several sites. Citizen groups responded with a site proposal of their own. The latter was found acceptable by the company. Thus informed consent was sought from and voluntarily given by those the project affected, and an acrimonious and protracted battle so common in other cases where a company has already invested heavily in decisions based on engineering studies alone was avoided (Borrelli, 36–39).

Study Questions

1 On June 5, 1976, Idaho's Teton Dam collapsed, killing eleven people and causing $400 million in damage. Drawing upon the concept of engineering as social experimentation, discuss the following facts uncovered by the General Accounting Office and reported in the press. (a) Because of the designers' confidence in the basic design of Teton Dam, it was believed that no significant water seepage would occur. Thus sufficient instrumentation to detect water erosion was not

installed. (b) Significant information suggesting the possibility of water seepage was acquired at the dam site 6 weeks before the collapse. It was sent through routine channels from the project supervisors to the designers, and arrived at the designers the day after the collapse. (c) During the important stage of filling the reservoir, there was no around-the-clock observation of the dam. As a result the leak was detected only 5 hours before the collapse. Even then the main outlet could not be opened to prevent the collapse because a contractor was behind schedule in completing the outlet structure (Shaw, 3; Boffee, 270–272).

2 The University of California uses tax dollars to develop farm machinery such as tomato, lettuce, and melon harvesters, fruit tree shakers, etc. Such machinery reduces the need for farm labor and raises farm productivity. It definitely benefits the growers. It is also said to benefit all of society. Farm workers, however, claim that replacing an adequate and willing work force by machines will generate social costs not offset by higher productivity. Among the costs they cite are the need to retrain farm workers for other jobs and the loss of small farms. Discuss if and how continuing farm mechanization may be viewed as an experiment.

3 Apply the social experimentation model to the DC-10 case described in Chap. 2. Specifically, in order to facilitate informed consent concerning dangers entailed by the plane's design, should Dan Applegate have been allowed to convey information to public representatives (in government or consumer groups) or directly (via newspapers) to the public who must decide whether to fly on DC-10 airplanes?

4 Models often influence thinking by effectively organizing and guiding reflection and crystallizing attitudes. Yet they usually have limitations and can themselves be misleading to some degree. Write a short essay in which you critically assess the strengths and weaknesses you see in the social experimentation model.

One possible criticism you might consider is whether the model focuses too much on the creation of new products, whereas a great deal of engineering involves the routine application of results from past work and projects. Another point to consider is how informed consent is to be measured in situations where groups are involved, as in the construction of a nuclear power plant near a community of people having mixed views about the advisability of constructing the plant.

5 Debates over responsibility for safety in regard to technological products often turn on whether the consumer should be considered mainly responsible ("buyer beware") or the manufacturer ("seller beware"). How might an emphasis on the idea of informed consent influence thinking about this question?

6 In the following passage from A Nation of Guinea Pigs, Marshall Shapo applies the concept of experimentation to the marketing of drugs. Comment on parallels and dissimilarities you see between the moral aspects of social experimentation in engineering and in drug marketing:

. . . experimentation is a label which connotes an attempt to solve problems in a fresh and novel way, using the subjects of the attempt as means to gather information. The image that the term conveys in the context of hazards involving products and processes tends to be a laboratory image. But much experimentation goes beyond the laboratory. In the process of testing and marketing new drugs, after procedures first limited to testing for toxicity and pharmacological effects, it takes place with increasingly large groups of patients in clinical trials. And although we do not conventionally attach the label "experimental" to the general marketing of products, it is clear that widespread distribution in fact involves a continuous process of experimentation. Especially with goods that are scientifically complex, the information-collecting goal of the experimenter is never attained in the formally investigational stages of the process. Some hazards may become apparent only after the products are used by millions of people, and over extended periods of time (Shapo, 30).

ENGINEERS AS RESPONSIBLE EXPERIMENTERS

What are the responsibilities of engineers to society? Viewing engineering as social experimentation does not by itself answer this question. For while engineers are the main technical enablers or facilitators, they are far from being the sole experimenters. Their responsibility is shared with management, the public, and others. Yet their expertise places them in a unique position to monitor projects, to identify risks, and to provide clients and the public with the information needed to make reasonable decisions.

The detailed content of engineer's responsibilities, in the sense of obligations, will be explored throughout the remainder of this book. At present we are interested in another of the senses of "responsibility" distinguished in Chap. 2. We want to know what is involved in displaying the virtue of being a responsible person while acting as an engineer. From the perspective of engineering as social experimentation, what are the general features of morally responsible engineers?

At least four elements are pertinent: a conscientious commitment to live by moral values, a comprehensive perspective, autonomy, and accountability (Haydon, 50–53). Or stated in greater detail as applied to engineering projects conceived as social experiments:

1 A primary obligation to protect the safety of and respect the right of consent of human subjects.

2 A constant awareness of the experimental nature of any project, imaginative forecasting of its possible side effects, and a reasonable effort to monitor them.

3 Autonomous, personal, involvement in all steps of a project.

4 Accepting accountability for the results of a project.

It is implied in the foregoing that engineers should also display technical competence and other attributes of professionalism. Inclusion of the above four requirements as part of engineering practice would then earmark a definite "style" of engineering. In elaborating upon this style, we will note some of the contemporary threats to it.

Conscientiousness

People act responsibly to the extent that they conscientiously commit themselves to live according to moral values. But moving beyond this truism leads immediately to controversy over the precise nature of those values. In Chap. 1 we adopted the minimal thesis that moral values transcend a consuming preoccupation with a narrowly conceived self-interest. Accordingly, individuals who think solely of their own good to the exclusion of the good of others are not moral agents. By *conscientious* moral commitment is meant a sensitivity to the full range of moral values and responsibilities that are relevant to a given situation, and the willingness to develop the skill and expend the effort needed to reach the best balance possible among those considerations.

The contemporary working conditions of engineers tend to narrow moral vision solely to the obligations that accompany employee status. As stated earlier, some 90 percent of engineers are salaried employees, most of whom work within large bureaucracies under great pressure to function smoothly within the organization. There are obvious benefits in terms of prudential self-interest and concern for one's family which make it easy to emphasize as primary the obligations to one's employer. Gradually the minimal negative duties, such as not falsifying data, not violating patent rights, and not breaching confidentiality, may come to be viewed as the full extent of moral aspiration.

Conceiving engineering as social experimentation restores the vision of engineers as guardians of the public interest, whose professional duty it is to guard the welfare and safety of those affected by engineering projects. And this helps to ensure that such safety and welfare will not be disregarded in the quest for new knowledge, the rush for profits, a narrow adherence to rules, or a concern over benefits for the many that ignores harm to the few.

The role of social guardian should not suggest that engineers force, paternalistically, their own views of the social good upon society. For, as with medical experimentation on humans, the social experimentation involved in engineering should be restricted by the participant's consent—voluntary and informed consent.

Relevant Information

Conscientiousness is blind without relevant factual information. Hence showing moral concern involves a commitment to obtain and properly assess all available information pertinent to meeting one's moral obligations. This means, as a first step, fully grasping the *context* of one's work which makes it *count as* an activity having a moral import.

For example, there is nothing wrong in itself with being concerned to design a good heat exchanger. But if I ignore the fact that the heat exchanger will be used as part of a still involved in the manufacture of a potent, illegal hallucinogen, I am showing a lack of moral concern. It is this requirement that one be aware of the wider implications of one's work which makes participation in, say, a super weapon design project morally problematic— and which makes it sometimes convenient for engineers self-deceivingly to ignore the wider context of their activities, a context that may rest uneasily with an active conscience.

Another way of blurring the context of one's work results from the ever increasing specialization and division of labor which makes it easy to think of someone else in the organization as responsible for what otherwise might be a bothersome personal problem. For example, a company may produce items with obsolescence built into them, or the items might promote unnecessary energy usage. It is easy to place the burden on the sales department: "Let *them* inform the customers—if the customers ask." It may be natural to thus rationalize one's neglect of safety or cost considerations, but it shows no moral concern.

These ways of losing perspective on the nature of one's work also hinder acquiring a full perspective along a second dimension of factual informa- tion: the *consequences* of what one does. And so while regarding engineer- ing as social experimentation points out the importance of context, it also urges the engineer to view his or her specialized activities in a project as part of a larger whole having a social impact—an impact that may involve a variety of unintended effects. Accordingly, it emphasizes the need for wide training in disciplines related to engineering and its results and the constant effort to imaginatively foresee dangers.

It might be said that the goal is to practice what Chauncey Starr once called "defensive engineering". Or perhaps more fundamental is the concept of "preventive technology" as described by Ruth Davis, who could have addressed the following lines equally well to engineers as she did to scientists and physicians:

> The solution to the problem is not in successive cures to successive science- caused problems; it is in their prevention. Unfortunately, cures for scientific ills are generally more interesting to scientists than is the prevention of those ills. We have the unhappy history of the medical community to show us the difficulties associated with trying to establish preventive medicine as a specialty.
>
> Scientists probably had more fun developing scientific defenses against nuclear weapons (that is, cures) than they would have had practicing preventive nuclear science during the development of the atomic bomb. Computer scientists find it more attractive to develop technological safeguards, after the fact, to prevent invasions of privacy associated with computer data banks than to develop good information practices along with the computer systems.
>
> However, it now seems quite clear that public patience with the cure always

following after the ill has worn thin. The public wants to see some preventive measures taken. Indeed, individuals have taken what can be called preventive technology into their own hands. We have seen the public in action in this way in its handling of the supersonic transport issue and its reaction toward siting of nuclear power plants. This is the reactive mode of practicing preventive technology, and it hinges on public recognition that technology is fallible (Davis, 213).

No amount of disciplined and imaginative foresight, however, can serve to anticipate all dangers. Because engineering projects are inherently experimental in nature, it is crucial for them to be monitored on an on-going basis from the time they are put into effect. While individual practitioners cannot conduct full-blown private environmental impact studies, they can choose to make the extra effort needed to keep in touch with the course of a project after it has officially left their hands. This is a mark of *personal* identification with one's work, a notion that leads to the next aspect of moral responsibility.

Moral Autonomy

People are morally autonomous when their moral conduct and principles of action are their *own,* in a special sense deriving from Kant. That, is, moral beliefs and attitudes must be held on the basis of critical reflection rather than merely through passive adoption of the particular conventions of one's society, church, or profession. This is often what is meant by "authenticity" in one's commitment to moral values.

Those beliefs and attitudes, moreover, must be integrated into the core of an individual's personality in a manner that leads to committed action. They cannot be agreed to abstractly and formally and adhered to merely verbally. Thus, just as one's principles are not passively imbibed from others when one is morally autonomous, so too one's actions are not treated as something alien and apart from oneself.

It is a comfortable illusion to think that in working for an employer, and thereby performing acts directly serving a company's interests, one is no longer morally and personally identified with one's actions. Selling one's labor and skills may make it seem that one has thereby disowned and forfeited power over one's actions (Lachs, 201–213).

Viewing engineering as social experimentation can help overcome this tendency and restore a sense of autonomous participation in one's work. As an experimenter, an engineer is exercising the sophisticated training that forms the core of his or her identity as a professional. Moreover, viewing an engineering project as an experiment that can result in unknown consequences should help inspire a critical and questioning attitude about the adequacy of current economic and safety standards. This also can lead to a greater sense of personal involvement with one's work.

The attitude of management plays a decisive role in how much moral autonomy engineers feel they have. It would be in the long-term interest of a high-technology firm to grant its engineers a great deal of latitude in exercising their professional judgment on moral issues relevant to their jobs. But the yardsticks by which a manager's performance is judged on a quarterly or yearly basis most often militate against this. This is particularly true in our age of conglomerates, when near-term profitability is more important than consistent quality and long-term retention of satisfied customers.

Accordingly engineers are compelled to look to their professional societies and other outside organizations for moral support. Yet it is no exaggeration to claim that the blue collar worker with union backing has greater leverage at present in exercising moral autonomy than do many employed professionals. A steel plant worker, for instance, who refused to dump oil into a river in an unauthorized manner was threatened with dismissal but his union saw to it that the threat was never carried out (Nader, 1972, 189). Or take the case of the automobile plant inspector who repeatedly warned his supervisors about poorly welded panels which allowed carbon monoxide from the exhaust to leak into the cab. Upon receiving no satisfactory response from the company he blew the whistle. The company wanted to fire him, but pressure from the union allowed him to keep his job. (The union, however, did not concern itself with the safety issue. It was probably as surprised as the company by the number of eventual fatalities traceable to the defect, the recall order those deaths necessitated, and the tremendous financial loss ultimately incurred by the company.) (Nader, 1972, 75–89)

Professional societies, originally organized as learned societies dedicated to the exchange of technical information, lack comparable power to protect their members, although most engineers have no other group to rely on for such protection. Only now is the need for moral and legal support of members in the exercise of their professional obligations being recognized by those societies.

Accountability

Finally, responsible people accept moral responsibility for their actions. Too often "accountable" is understood in the overly narrow sense of being culpable and blameworthy for misdeeds. But the term more properly refers to the general disposition of being willing to submit one's actions to moral scrutiny and be open and responsive to the assessments of others. It involves a willingness to present morally cogent reasons for one's conduct when called upon to do so in appropriate circumstances.

Submission to an employer's authority, or any authority for that matter, creates in many people a narrowed sense of accountability for the consequences of their actions. This was documented by some famous experiments conducted by Stanley Milgram during the 1960s (Milgram). Subjects would come to a laboratory believing they were to participate in a memory and learning test. In one variation two other people were involved, the "experimenter" and the "learner." The experimenter was regarded by the subject as an authority figure, representing the scientific community. He or she would give the subject orders to administer electric shocks to the "learner" whenever the latter failed in the memory test. The subject was told the shocks were to be increased in magnitude with each memory failure. All this, however, was a deception and a "setup". There were no real shocks and the apparent "learner" and the "experimenter" were merely acting parts in a ruse designed to see how far the unknowing experimental subject was willing to go in following orders from an authority figure.

The results were astounding. When the subjects were placed in an adjoining room separated from the "learner" by a shaded glass window, over half were willing to follow orders to the full extent: giving the maximum electric jolt of 450 volts. This was in spite of seeing the "learner" who was strapped in a chair, writhing in (apparent) agony. The same results occurred when the subjects were allowed to hear the (apparently) painful screams and protests of the "learner," screams and protests which became intense from 130 volts on. There was a striking difference, however, when subjects were placed in the same room within touching distance of the "learner." Then the number of subjects willing to continue to the maximum shock dropped by one-half.

Milgram explained these results by citing a strong psychological tendency in people to be willing to abandon personal accountability when placed under authority. He saw his subjects ascribing all initiative, and thereby all accountability, to what they viewed as legitimate authority. And he noted that the closer the physical proximity, the more difficult it becomes to divest oneself of personal accountability.

The divorce between causal influence and moral accountability is common in business and the professions, and engineering is no exception. Such a psychological schism is encouraged by several prominent features of contemporary engineering practice.

First, large scale engineering projects involve fragmentation of work. Each person makes only a small contribution to something much vaster. Moreover, the final product is often physically removed from one's immediate work place, creating the kind of "distancing" that Milgram identified as encouraging a lessened sense of personal accountability.

Second, corresponding to the fragmentation of work is a vast diffusion of accountability within large institutions. The often massive bureaucracies within which most engineers work are designed to diffuse and delimit areas of personal accountability within hierarchies of authority.

Third, there is frequently pressure to move on to a new project before the current one has been operating long enough to be observed carefully. This promotes a sense of being accountable only for meeting time schedules.

Fourth, the contagion of malpractice suits currently afflicting the medical profession is carrying over into engineering. With this comes a crippling preoccupation with legalities, a preoccupation which makes one wary of becoming morally involved in matters beyond one's strictly defined institutional role.

We do not mean to underestimate the very real difficulties these conditions pose for engineers who seek to act as morally accountable people on their jobs. Much less do we wish to say engineers are blameworthy for all the bad side effects of the projects they work on, even though they partially cause those effects simply by working on the projects. That would be to confuse accountability with blameworthiness, and also to confuse causal responsibility with moral responsibility. But we do claim that engineers who endorse the perspective of engineering as a social experiment will find it more difficult to divorce themselves psychologically from personal responsibility for their work. Such an attitude will deepen their awareness of how engineers daily cooperate in a risky enterprise in which they exercise their personal expertise toward goals they are especially qualified to attain, and for which they are also accountable.

Study Questions

1 A common excuse for carrying out a morally questionable project is, "If I don't do it somebody else will." This rationale may be tempting for engineers who typically work in situations where someone else might be ready to replace them on a project. Do you view it as a legitimate excuse for engaging in projects which might be unethical? (In your answer, comment upon the concept of responsible conduct developed in this section.)

2 Another commonly used phrase, "I only work here," implies that one is not personally accountable for the company rules since one does not make them. It also suggests that one wishes to restrict one's area of responsibility within tight bounds as defined by those rules (Lachs, 201–213). In light of the discussion in this section, respond to the potential implications of this phrase and the attitude represented by it when exhibited by engineers.

3 You have been asked to design an electronic vote counter for a legislative body. You have no difficulty with the physical features of the machine, but you begin to ask yourself some questions. If heretofore all votes, except for secret ballots, were by a show of hands, should a display board be provided indicating each individual vote? Or would total tallies be sufficient, thereby assuring anonymity of voting on each occasion? What would be the implication of each option for respecting the public's right to know? How much need you worry about unauthorized tampering with such a machine? Describe to what extent the model

of social experimentation can be applied to the introduction of the vote counter. (For a short report on the West German parliament's reluctance to put a vote counter to use in 1971, see the *Los Angeles Times* article by Joe Alex Morris, Jr., cited in the Bibliography.)

4 Threats to a sense of personal responsibility are neither unique to nor more acute for engineers than they are for others involved with engineering and its results. For in general, *public* accountability also tends to lessen as professional roles become narrowly differentiated. With this in mind, critique each of the remarks made in the following dialogue. Is it true, or partially true? What needs to be added to make it accurate?

Engineer: My responsibility is to receive directives and to create products within specifications set by others. The decision about what products to make and their general specifications are economic in nature and made by management.
Scientist: My responsibility is to gain knowledge. How the knowledge is applied is an economic decision made by management, or else a political decision made by elected representatives in government.
Manager: My responsibility is solely to make profits for stockholders.
Stockholder: I invest my money for the purpose of making a profit. It is up to managers to make decisions about the directions of technological development.
Consumer: My responsibility is to my family. Government should make sure corporations do not harm me with dangerous products, harmful side effects of technology, or dishonest claims.
Government regulator: By current reckoning, government has strangled the economy through overregulation of business. Accordingly at present on my job, especially given decreasing budget allotments, I must back off from the idea that business should be policed, and urge corporations to assume greater public responsibility.

THE BART CASE

Modern railroading as exemplified in the Bay Area Rapid Transit System (BART), the suburban commuter rail system that links San Francisco with the cities across its bay, can be used to illustrate the experimental nature of engineering and the importance of engineers and managers showing moral responsibility when working on such projects.

Background

The pioneering years of railroading indicate that technological experimentation is usually highly fruitful. For example, early fears about the effects of high-speed travel—of sparks showering the countryside, of animals being frightened by the noise and fast movement—were proven to be unfounded. The benefits to agriculture, industry, and commerce, moreover, were immense. And society learned that to secure those benefits it could live with the loss of forests to railroad ties and fuel or the cycle of settlement building and abandonment entailed by the construction of new railroads.

As technological innovation in railroading accelerated, however, the trend to do the fashionable thing for its own sake increasingly predominated. For example, railroads took over in instances where common sense would have dictated the continued use of barges on canals. To some extent BART is a recent example of that trend. Developed to incorporate the latest "space age" technology in its design, it ended up as more expensive and less reliable than its traditional counterparts.

BART was built with tax funds, and its construction was characterized by tremendous cost overruns and numerous delays. Much of this can be ascribed to the introduction of innovative methods of communicating with individual trains and of controlling them automatically. In addition, plain fail-safe operation was replaced by complex redundancy schemes. (Fail-safe features simply cause a train to stop if something breaks down; redundancy features try to keep trains running by switching the faulted components to alternate ones.) The rationale given for this approach was that the system could be sold to the public only if it involved glamorous and exciting gadgetry.

Some of the technical aspects of BART are described in articles which appeared in the journal *IEEE Spectrum* (Friedlander, March 1973 and April 1973). A full-length book on some of the ethical and political issues connected with BART (Anderson et al.) will be discussed later.

Responsibility and Experimentation

The opportunity to build a rail system from scratch, unfettered by old technology, was a challenge which excited many engineers and engineering firms. Indeed, altogether the project was an interesting experiment. Yet among the engineers who worked on it were some who came to feel that too much experimentation was going on without proper safeguards. Safety features were given insufficient attention and quality control was poor, they thought.

Three engineers in particular—Holger Hjortsvang, Robert Bruder, and Max Blankenzee—identified dangers that were to be recognized by management only much later. They saw that the automatic train control was unsafely designed. Moreover, schedules for testing it and providing operator training prior to its public use were inadequate. Computer software problems continued to plague the system. Finally, there was insufficient monitoring of the work of the various contractors hired to design and construct the railroad. These inadequacies were to become the main causes of several early accidents.

The three wrote a number of memos and voiced their concerns to their employers and colleagues. Their initial efforts were directed within organizational channels to both their immediate supervisors and the next two higher levels of management, but to no avail. Yet they refused to wait passively for accidents to occur, and resolved to do more.

Up to this point Hjortsvang, Bruder, and Blankenzee clearly displayed the kind of moral responsibility described in the last section. They were conscientious in refusing to lose sight of their primary obligation to the public—that is, their obligation to what was, in effect, the "subject " of this particular engineering "experiment." They were imaginative in foreseeing dangers. They were personally and autonomously involved. And they were willing to accept moral accountability for their participation in the project.

Of special interest in the case is that for the most part the three engineers were not specifically assigned or authorized by the BART organization to check into the safety of the automatic control system. Hjortsvang, for example, first identified the dangers when he was sent to Westinghouse (a BART subcontractor) primarily to observe, not supervise, the development of the control system. Similarly, Robert Bruder worked for the construction department, not the operations department which had responsibility for the train control. Thus, both engineers looked to the wider implications of the specific tasks assigned them within the organization. They refused to have their moral responsibility circumscribed within a narrow organizational bailiwick.

Controversy

The controversial events that followed as the engineers sought to pursue their concerns further are described and interpreted from the opposing viewpoints of the engineers and management (and others) in the book *Divided Loyalties* by Robert M. Anderson et al. (cited in the Bibliography). Here we can mention only two or three of them.

Hjortsvang wrote an anonymous memo summarizing the problems, and distributed copies of it to nearly all levels of management, including the project's general manager. The memo argued that a new systems engineering department was needed, a department that Hjortsvang had also requested in an earlier signed memo. Distribution of such an unsigned memo was regarded by management as suspicious and unprofessional since it was done outside the normal channels of accountability within the organization. Later, when its author was identified, management decided Hjortsvang was motivated by self-interest and power-seeking since it could be assumed that he wished to become the head of such a department.

Next, the three engineers contacted several members of BART's board of directors when their concerns were not taken seriously by lower levels of management, thereby departing again from formally approved channels. In order to obtain an independent view, they also contacted a private engineering consultant who on his own wrote an evaluation of the automatic train control. One of the directors, Dan Helix, listened sympathetically and agreed to contact top management while keeping the engineers' names confidential. But to the shock of the three, Helix released copies of their unsigned memos and the consultant's document to the local newspapers.

Aftermath

Management immediately sought to locate the source of Helix's information. Fearing reprisals, the engineers at first lied to their supervisors and denied their involvement. At Helix's request, however, they later agreed to reveal themselves by going before the full board of directors in order to seek a remedy for the safety problems. On that occasion they were unable to convince the board of those problems. One week later they were fired. The grounds given for the dismissal were insubordination, incompetence, lying to their superiors, causing staff disruptions, and failing to follow understood organizational procedures.

Two years later they sued BART for damages, and settled out of court. They were assisted in their court case by an *amicus curiae* ("friend of the court") brief filed by the Institute of Electrical and Electronics Engineers (IEEE). This legal brief noted in their defense that it is part of each engineer's professional duty to promote the public welfare, as stated in IEEE's code of ethics. In 1978 IEEE presented them each with its Award for Outstanding Service in the Public Interest for "courageously adhering to the letter and spirit of the IEEE code of ethics."

Comments

We will leave to our readers the task of assessing the extent to which the three engineers and BART's management acted responsibly (see Study Questions below). The complexities revealed in *Divided Loyalties*—the first full-scale scientific study of an instance of whistle-blowing in engineering—show the case is hardly a simple one. Here we wish to comment upon two attitudes held by the authors of that book, attitudes germane to the topic of moral responsibility and deserving of mention because of the frequency with which similar arguments are heard in other contexts.

The authors' final verdict is that the BART case "can be viewed as not really involving safety or ethics to any marked degree" (Anderson et al., 353). We disagree. The main basis for that verdict seems to be the claim that BART's complex organizational structure alone was to blame for the conflicts which helped precipitate the incidents. For example, the engineers were given considerable freedom to determine for themselves the specific tasks they were to pursue, but granted little authority to implement changes they felt were needed. Frustration on their part was therefore to be expected.

This, however, fails to show that ethical issues were not involved. On the contrary, it shows how ethical issues can arise out of problems associated with organizational structure. Indeed, the conflicts engendered by the social, political, and economic settings of an organization quite frequently form the background for the ethical problems engineers confront when concerned about how best to ensure the safety of their projects.

The authors' verdict may also have resulted from a lack of clarity about what an ethical problem is. For they emphasize that there were *no villains* in the BART episode. Those involved were basically good people trying in the main to do their jobs responsibly even if they were influenced to some degree by self-interest. This seems to imply that ethical situations must always involve bad people who are opposed by good people—a melodramatic view of morality. Yet surely the question of how best to assure safety in any engineering project is a moral issue, whatever the ultimate personal motivations of the people involved in it. Ethics can involve a decision between good and better just as much as a conflict between good and bad.

Study Questions

1 The authors of *Divided Loyalties* conclude, without argument that "the engineers took a narrow and technical view of the issues which disturbed them" (Anderson et al., 351). Present and defend your view as to whether, and in what respects, the BART engineers and BART management acted correctly. In doing so, discuss alternative courses of action that either or both groups might have pursued.

2 The authors of *Divided Loyalties* also state that the three engineers suffered from a "political naivete which permitted them to carry their grievance as far as they did" (Anderson et al., 351). BART was a public project. In what sense, if any, should political considerations have entered into the decisions made by the three engineers?

3 General attitudes toward possibilities and risks can greatly influence engineering practice. They include Pollyanna optimism ("It must go our way"), cautious optimism ("Think positive, but be careful"), tougher optimism ("We have a good chance, but should be ready for the worst"), pessimism ("We have a chance, but should expect the worst to happen"), despair ("We haven't got a prayer"), and so on. Another attitude is referred to as "technofix": for any problem created by technology, we can find a technological solution. Working with several other

students, write a short dialogue concerning engineering as social experimentation in which various characters argue for and against several of these attitudes.

CODES OF ETHICS

Engineering societies have followed other professional groups in developing official codes of professional ethical conduct. These codes serve a number of functions, which helps explain why they have been given so much attention. The following are some of the more prominent of those functions, prominent with both their positive and negative consequences. It is suggested that in reading this section the reader examine the sample codes of ethics given in the Appendix.

Roles of Codes

Inspiration and Guidance   Codes provide a positive stimulus for ethical conduct and helpful guidance and advice concerning the main obligations of engineers. Often they succeed in inspiring by using language with positive overtones. This can introduce a large element of vagueness, as in phrases like "safeguard the public safety, health, and welfare," a vagueness which may lessen their ability to give concrete guidance.

Since codes are brief documents, they give rather general guidelines. More specific guidance may be given in supplementary statements which tell how to apply the code, such as the "Suggested Guidelines" formerly set forth by the Engineers Council for Professional Development. Further specificity may also be attained by the interpretation of codes. This is done for engineers by the National Society of Professional Engineers. It has established a Board of Ethical Review which applies the Society's code to specific cases and publishes the results in *Professional Engineer* and periodic volumes entitled *NSPE Opinions of the Board of Ethical Review.*

Support   Codes give positive support to those seeking to act ethically. A publicly proclaimed code allows an engineer who is under pressure to act unethically to say: "I am bound by the code of ethics of my profession, which states that. . . ." This by itself gives engineers some group backing in taking stands on moral issues. Moreover, as in the BART case, codes can potentially serve as legal support in courts of law for engineers seeking to meet work-related moral obligations.

Deterrence and Discipline   Codes can serve as the formal basis for investigating unethical conduct. Where such investigation is possible, a prudential motive for not acting immorally is provided as a deterrent. This generally requires paralegal proceedings designed to get at the truth about a given charge without violating the personal rights of those being investigated. In the past, engineering professional societies have been reluctant to undertake such proceedings because they have lacked the appropriate sanctions needed for punishment of misconduct. Unlike the American Bar Association and some other professional groups, engineering societies cannot revoke the right to practice engineering in this country. Yet the American Society of Civil Engineers, for example, does currently suspend or expel members whose professional conduct has been proven unethical, and this alone can be a powerful sanction when combined with the loss of respect from colleagues and the local community that such action is bound to produce.

Education and Mutual Understanding   Codes can be used in the classroom and elsewhere to prompt discussion and reflection on moral issues and to encourage a shared understanding among professionals, the public, and government organizations concerning the moral responsibilities of engineers. They can help do this because they are widely circulated and officially approved by professional societies.

Contributing to the Profession's Public Image   Codes can present a positive image to the public of an ethically committed profession. Where the image is warranted, it can help engineers more effectively serve the public. It can also win greater powers of self-regulation for the profession itself while lessening the demand for more government regulation. Where unwarranted, it reduces to a kind of false window-dressing that ultimately increases public cynicism about the profession.

Protecting the Status Quo   Codes establish ethical conventions, which can help promote an agreed-upon minimum level of ethical conduct. But it can also stifle dissent within the profession. On occasion this has positively discouraged moral conduct and caused serious harm to those seeking to serve the public. In 1932, for example, two engineers were expelled from the American Society of Civil Engineers for violating a section of its code forbidding public remarks critical of other engineers. Yet the actions of those engineers were essential in uncovering a major bribery scandal related to the construction of a dam for Los Angeles County (Layton, 1980, 26).

Promoting Business Interests   Codes can place "restraints of commerce" on business dealings with primary benefit to those within the profession. Basically self-serving items which enter into codes can take on great undue influence. Obviously there is disagreement about which, if any, entries function in these ways. Some engineers believe that in the past the codes were justified in forbidding competitive bidding, while others agree with the decision of the Supreme Court in the case of the National Society of Professional Engineers vs. the United States (April 25, 1978) that such a restriction is inappropriate.

Codes and the Experimental Nature of Engineering

Given that codes may play all these roles, which functions are the most valuable and should be emphasized and encouraged? This is an important question, if only because its answer can greatly influence the very wording of codes. For example, if the disciplinary function is to be emphasized, every effort would have to be made to ensure clear-cut and enforceable rules. This would also tend to make statements of minimal duty predominant, as with standards and laws, rather than statements concerned with higher ideals. By contrast, if the emphasis is to be on inspiration, then statements of high ideal might predominate. Nothing is less inspirational than arid, legalistic wordings, and nothing is less precise than highly emotive exhortations.

The perspective of engineering as social experimentation provides some help in deciding which functions should be primary in engineering codes. It clearly emphasizes those which best enable concerned engineers to express their views freely—especially about safety—to those affected by engineering projects. Only thus can clients and the public be educated adequately enough to make informed decisions about such projects. But as we have already noted and will discuss in more detail later, contemporary working conditions within large corporations do not always encourage this freedom of speech—conditions for which a code of ethics can provide an important counterbalance. Thus the supportive function seems to us of primary importance.

The guidance, inspirational, and educational functions of engineering codes are important also, as is their role in promoting mutual understanding among those affected by them. In seeking to create a common understanding, however, code writers must take every precaution to allow room for reasonable differences among individuals. Wordings in past codes, for example, sometimes used religious language not acceptable to many who did not share that orientation. Codes, we must bear in mind, seek to capture the essential substance of professional ethics; they can hardly be expected to express the full moral perspective of every individual.

The disciplinary function of engineering codes is in our view of secondary importance. There are scoundrels in engineering, as there are everywhere. But when exposed as such, they generally fall subject to the law. Developing elaborate paralegal procedures within professional societies runs the risk of needlessly and at considerable cost duplicating a function better left to the real legal system. At most, enforcement of professional ethics by professional societies should center upon areas not covered by law which can be made explicit and clear-cut, preferably in separate code sections specifically devoted to those areas. In any case, the vast majority of engineers can be counted on to act responsibly in moral situations unless *discouraged* from doing so by outright threats and lack of support on the part of the employers.

Probably the worst abuse of engineering codes in the past has been to restrict honest moral effort on the part of individual engineers in the name of preserving the profession's public image and protecting the status quo. Preoccupation with keeping a shiny public image may silence the healthy dialogue and lively criticism needed to ensure the public's right to an open expression. And an excessive interest in protecting the status quo may lead to a distrust of the engineering profession on the part of both government and the public. The best way to *increase* trust by the public and government is by encouraging and aiding engineers to speak freely and responsibly about the public safety and good as they see it. And this includes a tolerance for criticisms of the codes themselves. Perhaps the worst thing that can happen is for codes to become "sacred documents" that have to be accepted uncritically.

Limitations of Codes

Most codes are limited in several major ways. Those limitations restrict codes to providing only very general guidance, which in turn makes it essential for engineers to exercise a personal moral responsibility in their role as social experimenters rather than expect codes to solve their moral problems by serving as simple algorithms. The limitations of codes are as follows.

First, as we have already mentioned, codes are restricted to general and vague wording. Because of this they are not straightforwardly applicable to all situations. After all, it is not humanly possible to foresee the full diversity of moral problems that can arise within a complex profession like engineering. New technical developments and shifting social and organizational structures combine to generate continually new and often unpredictable conditions. And even in the case of foreseeable situations it is not possible to word a code so that it will apply in every instance. Attempting to do so would yield something comparable to the intricate set of laws governing engineering rather than a manageable code.

A sense of responsibility is indispensable for the skillful and at times creative application of code guidelines to concrete situations. It is also the only way certain abuses of codes can be avoided—for example, abuses such as special interpretations being placed on general entries, or legalistic glosses on specific entries, to serve the private gain or convenience of specific individuals or groups.

Second, it is easy for different entries in codes to come into conflict with each other. Usually codes provide no guidance as to which entry should have priority in those cases, thereby creating moral dilemmas.

For example, take the following two former entries from the National Society of Professional Engineers (NSPE) code. Section 1: "The Engineer will be guided in all his professional relations by the highest standards of integrity, and will act in professional matters for each client or employer as a faithful agent or trustee." Section 2: "The Engineer will have proper regard for the safety, health, and welfare of the public in the performance of his professional duties." Which was the more applicable in the DC-10 case mentioned in Chap. 2, where an engineer was told to ignore a situation he believed threatening to the public safety on the basis of a business decision made in the interests of his company?

Recent codes have attempted to address this important area of potential conflict. The NSPE code now states: "Engineers shall hold paramount the safety, health, and welfare of the public in the performance of their professional duties." The word "paramount" means "most important or superior in rank." But even so it is unclear that the provision means engineers should never, under any circumstances, follow a client's or company's directives because they believe those directives might not serve the best interests of the public. This is an issue we will return to later in part III of our book. But here we emphasize again the need for responsible engineers who are able to make reasonable assessments of what "paramount" amounts to in cases where two professional obligations conflict.

A third limitation on codes is that they cannot serve as the final moral authority for professional conduct (Ladd, 1980, 154). To accept the current code of a professional society as the last moral word, however officially endorsed it may be, would be to lapse into a type of ethical conventionalism. It will be recalled that ethical conventionalism is the view that a particular set of conventions, customs, or laws is self-certifying and not to be questioned simply because it is the set in force at a given time or for a given place. Such a view, of course, rules out the possibility of criticizing that set of conventions from a wider moral framework.

Consider once again the following entry in the pre-1979 versions of the NSPE code: "He [the engineer] shall not solicit or submit engineering proposals on the basis of competitive bidding." This prohibition was felt by the NSPE to best protect the public safety by discouraging cheap engineering proposals which might slight safety costs in order to win a contract. Critics of the prohibition, however, contended that it mostly served the self-interest of engineering firms and actually hurt the public by preventing the lower prices that might result from greater competition. In a 1978 decision, the National Society of Professional Engineers vs. the United States, the Supreme Court ruled that the ban on competitive bidding was unconstitutional and not appropriate in a code of ethics.

The point here is not who holds the correct moral view on this issue—that is a matter of on-going debate and discussion. And indeed, it is precisely our point that no pronouncement by a code current at any given time should ever be taken as the final word silencing such healthy debates. Codes, after all, represent a compromise among differing judgments, sometimes developed amidst heated committee disagreements. As such, they have a great "signpost" value in suggesting paths through what can be a bewildering terrain of moral decision making. But equally as such they should never be treated as "sacred canon."

The fourth limitation on codes is their proliferation. Andrew Oldenquist (a philosopher) and Edward Slowter (an engineer and former NSPE president) point out how the existence of separate codes for different professional engineering societies can give members the feeling that ethical conduct is more "relative" than it is, and how it can convey to the public the view that none of the codes is "really right." The societies themselves have attempted regularly to establish a unified code of ethics, so far to no avail. But Oldenquist and Slowter have also demonstrated the substantial agreement to be found among the various engineering codes. They summarize the core concepts in each and arrange them in order of significance as having to do with (1) the public interest, (2) qualities of truth, honesty, and fairness, and (3) professional performance. And they emphasize that the time appears to be right for a unified code (Oldenquist and Slowter, 1979, 8–11). Perhaps sometime soon this limitation will no longer exist.

Study Questions

1 Apply a code of ethics taken from the Appendix—or all of those codes—to the short cases presented as study questions at the end of the section entitled "Normative, Conceptual, and Descriptive Inquiries" in Chap. 1. Discuss the possible effectiveness of the code(s) as a deterrence to unethical behavior in these cases.

2 Comment on the following passage: "A code only sets the limits beyond which behavior will be condemned, and the moral level is not high when all or most of those who live under it always act within a hairline of those limits. Codes, in fact, are for criminals and competitors, not for professions that want to be known as dedicated" (Barzun, 67). Specifically, is this true of the engineering codes given in the Appendix?

3 Respond to the following claim: "Even if substantial agreement could be reached on ethical principles and they could be set out in a code, the attempt to impose such principles on others in the guise of ethics contradicts the notion of ethics itself, which presumes that persons are autonomous moral agents" (Ladd, 1980, 154). Is the idea of an officially prescribed, authoritative code of ethics somehow incompatible with an appreciation of the importance of moral autonomy in individuals?

4 Closely examine one of the codes given in the Appendix. Are there any entries in it which you think should not be there? Why? Are there any important omissions you see in the code?

5 Discuss the advertisement reprinted below in light of your understanding of engineering codes of ethics.

A code of ethics isn't something you post on the bulletin board.

It's something you live every day.

Suddenly everybody seems to be rediscovering ethics.

In the business community, in Congress, on the campus and in the pulpit.

We think the trend is healthy. And needed. So we'd like to disclose a discovery of our own on this subject.

We found a long time ago that when it comes to any sort of corporate decree, the more you reduce it to writing the more you reduce participation.

It's much better, we learned, to create a working environment in which communication is a two-way process. And corporate goals are shared.

So that your code of ethics is expressed not in a news release, but in the release of appropriate thought and action.

Nobody's perfect, but it seems to work.

As our chairman put it: "The character of this company is simply a reflection of how Pennwalt people think and act. *That's* our code of ethics."

And so it is.

Admittedly, it's an approach that places more stress on the integrity and good judgment of our people than on manuals from Personnel. (A *lot* more stress.)

But it pays off. In pride. In performance. In a belief that the work we do is important. And in the enhancement of our worldwide reputation.

You might say it's the difference between a bulletin that goes up on the board, and the life that goes on every day.

(We have a brief booklet on corporate citizenship which we believe covers this subject. If you'd like one, just write our Director of Corporate Communications.)

Pennwalt Corporation, Three Parkway, Philadelphia, Pa. 19102.

For 126 years we've been making things people need—including profits.

PENNWALT
CHEMICALS ■ EQUIPMENT
HEALTH PRODUCTS

Courtesy of Penwalt Corporation.

6 Consider the following two entries in the 1981 Code of the National Society of Professional Engineers. (a) "Engineers shall cooperate in extending the effectiveness of the profession by interchanging information and experience with other engineers and students." (b) "Engineers shall not disclose confidential information concerning the business affairs or technical processes of any present or former client or employer without his consent." Suppose that the two entries come into conflict, such as when improving the knowledge and skill of another engineer or student might best be done by passing on confidential information. Which entry should take precedence, and why? Do you think the code should be modified so as to explicitly state which entry should take precedence?

A BALANCED OUTLOOK ON LAW

The 1969 Santa Barbara offshore spill of 235,000 gallons of crude oil blackened 30 miles of spectacular beaches, damaged wildlife, and hurt the local tourist trade. Predictably, the disaster prompted demands for new laws and tighter controls to prevent such occurrences in the future (Lawless, 233–247). A group of Southern Californians staged a burning of gasoline credit cards issued by the offending oil company, Union Oil, only to be taken to task by a local newspaper for taking the wrong aim. The newspaper argued that the gas station operators, who would suffer the most from a boycott, were not at fault. The real offenders were the federal authorities who required less stringent safeguards in offshore drilling than did California state authorities, it was claimed.

Yet we might well ask, who would be involved in drafting safety regulations for offshore drilling? Obviously experienced petroleum engineers, geologists, and well drillers, members of the same group which had prepared the state regulations and who—in their capacity as oil company employees—had also conducted the drilling off the coast of Santa Barbara. If expert knowledge was available, then why was it not applied, law or no law?

It is worth noting that some safeguards were indeed required by federal law. Following the Santa Barbara incident, Secretary of the Interior Walter Hickel ordered an inspection of the thousands of offshore oil wells, mostly in the Gulf of Mexico. The inspection showed that hundreds lacked mandatory safety chokes. Hickel ordered prosecutions and later justified his tough approach to pollution with what has been called "Hickel's law": "You've got to hit them [i.e., polluters] with a two-by-four to make them believe you" (Rosenbaum, 129).

Is it really necessary to burden engineering practice with ever more—and increasingly restrictive—rules? Earlier we discussed the bases for responsible action. Here we shall examine the role of formal rules and their ethical implications. The model of engineering as social experimentation will assist us again as we consider the interaction of rules with the engineering process. The problem of product liability and safe design will be postponed until we take up a more detailed analysis of risk in Chap. 4.

A Regulated Society

In order to live, work, and play together in harmony as a society, we need to carefully balance individual needs and desires against collective needs and desires. Ethical conduct, which by definition includes a strong element of altruism, provides such a balance. Unfortunately people all too frequently disagree on what constitutes right action in specific instances, even when they agree on ultimate goals. At such times we need to negotiate, and if a compromise can be agreed upon, it should be recorded for repeated reference and use.

Engineers can play an active role in establishing or changing rules as well as in enforcing them. Indeed, some people would say that the engineer's ethical duties should be limited to just such activities—in addition to following accepted rules of conduct, of course (Florman, 1978, 30–33).

At various times in history, and in various countries, engineers have had less say in how rules affecting their work were made or carried out, except perhaps for a few who were among a ruler's trusted advisors; often engineers were merely subject to those rules. We assume that the time of Hammurabi fits this description.

1758 B.C.: Babylon's Building Code   Hammurabi as King of Babylon was concerned with strict order in his realm, and he decided that the builders of his time should also be governed by his laws. Thus he decided as follows:

> If a builder has built a house for a man and has not made his work sound, and the house which he has built has fallen down and so caused the death of the householder, that builder shall be put to death. If it causes the death of the householder's son, they shall put that builder's son to death. If it causes the death of the householder's slave, he shall give slave for slave to the householder. If it destroys property he shall replace anything it has destroyed; and because he has not made sound the house which he has built and it has fallen down, he shall rebuild the house which has fallen down from his own property. If a builder has built a house for a man and does not make his work perfect and the wall bulges, that builder shall put that wall into sound condition at his own cost (Hammurabi; also quoted by Firmage, 7).

The substantive or normative part of Babylon's building code is admirable for its succinct statement. The procedural aspects would find little approval today, although we cannot help but wistfully reflect on how little of a bureaucracy it probably took to maintain standards. One can imagine how builders passed on their carefully drawn rules for sound design from generation to generation. There was indeed a powerful incentive for self-regulation! In other words, the law was broad and the specifics of how to comply with it were left to those presumably best able to formulate them for each application—the builders themselves. We might note that this was by no means a simple matter, for "the Babylonians found only deep alluvium in their flood plains between the Tigris and the Euphrates, which settled under the weight of their cities" (Sowers, 389).

Let us turn to another example, some four milennia later, where procedural aspects and regulations in detail took the form of ready-made standards.

1852 A.D.: The U.S. Steamboat Code   Early steam engines were large and cumbersome. So to make them more practical for use, James Watt, and later Oliver Evans and Richard Trevethick, increased steam pressure, did away with the condenser in some models, and thus ushered in the age of compact, portable sources of motive power. In spite of these pioneers' careful calculations and guidelines, however, boiler explosions were frequent, particularly on steamboats. Nowhere was the problem as acute as in the United States, where riverboats were vying for trade on the great midwestern rivers. Races were common, boilers were stressed beyond their limits, and safety valves were disabled to keep steam pressure up; 233 explosions contributed to a total of 2563 persons killed and 2097 injured during the period 1816 to 1848. One explosion alone on the "Moselle" in Cincinnati in 1838 claimed 151 lives (Burke).

Demands for safety rules finally moved Congress to exert its river and interstate regulatory powers. Steamboat interests objected. It was argued that the prudent self-interest of steamboat owners and operators would in itself dictate caution. Cumbersome rules and an unyielding bureaucracy were predicted. But the self-regulatory commitments by owners and operators were clearly not in evidence. So in 1838 a law was passed which provided for the inspection of the safety features of ships and their boilers and engines. The occurrence of an explosion was to be taken as prima facie evidence of negligence and any loss of life was to be considered manslaughter.

But the 1838 law turned out to be ineffective. Ship owners could find corruptible inspectors. Even honest inspectors were not much help. They had no training and the law did not specify how a safety check should be conducted. Nevertheless, after a safety check was carried out, a ship owner could claim to be blameless. Boiler explosions continued unabated.

Among those who were troubled by the situation was Alfred Guthrie, an engineer from Illinois. Guthrie had inspected, at his own expense, about

200 steamboats to learn the causes of boiler explosions and written a report on his findings. His recommendations were published by a Senator Shields of Illinois and included in Senate documents. By 1852, when a new steamboat bill came before Congress, the groundwork had been carefully laid and an effective law was passed. Guthrie was made the first supervisor of the regulatory agency established by the law.

Congress was able to intervene as it did because of its powers to regulate interstate shipping. But even then it was left to ad hoc associations, insurance companies, and later the American Society of Mechanical Engineers to promulgate the standards which would govern the manufacture of steam boilers and their operation in mines, factories, and railroads. In France, boiler safety standards were earlier and more rapidly promulgated under the more centralized state authority of the Napoleonic code. Between 1823 and 1830 a committee of engineers, assisted by prominent scientists of the time, developed accurate steam tables, stress values for metals, and design standards which called for hemispherical end plates and initial testing of boilers at three times their expected operating pressure. France had very few boiler explosions thereafter, as did the United States after 1852.

The Trend Toward Greater Detail

In Hammurabi's time one could let the law take care of building failures *after* the structure had failed. While many houses may have crumbled, there were probably not many casualties associated with any one occasion (unless an earthquake had struck). However, when 150 passengers and crew members can be killed all at once by a boiler explosion and the ship is likely to sink, there will be demands for rules which prevent such accidents from happening in the first place. As technology's machines became more complex, simplicity in rule-making appeared to be doomed. The 1852 steamboat law even had to regulate the qualifications of steamboat inspectors.

But lawmakers cannot be expected always to keep up with technological development. Nor would we necessarily want to see laws changed with each new innovation. What is needed are regulating agencies and commissions—the Food and Drug Adminstration (FDA), Federal Aviation Agency (FAA), and the Environmental Protection Agency (EPA) are examples of such in the United States—to fill the void. They employ experts who can set up precise regulations. And even though they are independent and belong to neither the judicial nor the executive branches of government, their rules have, for all practical purposes, the effect of law.

Industry tends to complain that excessive restrictions are imposed on it by regulatory agencies. But then one needs to reflect on why regulations may have been necessary in the first place. Take, for example, the U.S. Consumer Product Safety Commission's rule for baby cribs which specifies that "the distance between components (such as slats, spindles, crib rods, and corner posts) shall not be greater than 2 3/8 inches at any point." This rule came about because some manufacturers of baby furniture had neglected to consider the danger of babies strangling in cribs or had neglected to measure the size of babies' heads (Lowrance, 134).

Again, why must regulations be so specific when broad statements would appear to make more sense? When the EPA adopted rules for asbestos emissions in 1971, it was recognized that strict numerical standards would be impossible to promulgate. Asbestos dispersal and intake, for example, are difficult to measure. So, being reasonable, EPA specified a set of work practices to keep emissions to a minimum—that asbestos should be wet down before handling, for example, and disposed of carefully.

> [A wrecking company], after promising repeatedly to comply with the rules, came along and demolished buildings without taking any of the precautions— thereby endangering its workers and the surrounding community. The violations were so blatant, EPA felt, and civil procedures so inadequate under the Clean Air Act, that the agency asked for and received a criminal indictment. . . . The U.S. Supreme Court overruled the Court of Appeals . . . and threw out the charges. Thanks to the High Court's technical illiteracy, EPA might now be justified in attempting to prescribe voluminous measurement techniques covering all possible asbestos-generating situations since its reasonableness led to an all but unenforceable rule. The engineering community would then snicker and joke about EPA's foolishness. Wouldn't it be better for the construction industry to police itself, and for demolition instructions with regard to asbestos to be clearly specified by contract (S. Ross, 6)?†

Industrial Standards

There is one area in which industry usually welcomes greater specificity, and that is in regard to standards. Standards facilitate the interchange of components, they serve as ready-made substitutes for lengthy design specifications, and they decrease production costs.

Standards consist of explicit specifications which, when followed with care, assure that stated criteria for interchangeability and quality will be attained. Examples range from automobile tire sizes to computer languages. Table 3-1 lists purposes of standards and gives some examples to illustrate those purposes.

Standards are established by companies for in-house use, are adopted by professional associations and trade associations for industry-wide use, and

---

†Modifications in the Clean Air Act eventually permitted EPA to issue enforceable work practice rules (private communication from S. Ross).

TABLE 3-1
TYPES OF STANDARDS

Criterion	Purpose	Selected examples
Uniformity of size and physical properties	Accurate measurements; interchangeability	Standards of weights and volumes; screw thread dimensions; film size
Safe and economic operation	Prevention of deaths and injuries and reduction of loss of income due to faulty operation	Vehicle speed and loading capacities; procedures for inspections and tests
Quality of personnel	Competency in carrying out tasks	Engineering school accreditation; professional engineer's examination
Use of accepted design methods	Sound design, ease of communication	Drawing standards and symbols; calculation of heat loss
Product safety and reliability	Prevention of deaths, injuries, and loss of property	Electrical code; pressure vessel standards
Quality of end product	Fair value for price	Plywood grade; lamp life

may also be prescribed as parts of laws and official regulations. The latter would be examples of mandatory standards, which frequently arise from lack of adherence to voluntary standards.

Standards help not only the manufacturers; they also benefit the client and the public. They preserve some competitiveness in industry by reducing overemphasis on name brands and giving the smaller manufacturer a chance to compete. They assure a measure of quality and thus facilitate more realistic tradeoff decisions.

Standards can also be a hindrance at times. For many years they were mostly descriptive, specifying, for instance, how many joists of what size should support a given type of floor. Clearly such standards tended to stifle innovation. The move to performance standards, which in the case of a floor may merely specify the required load-bearing capacity, has alleviated that problem somewhat. But other difficulties can arise when special interests (manufacturers, trade unions, exporters/importers, etc.) manage to impose unnecessary provisions on standards, or remove important provisions from them, to secure their own narrow self-interest. Requiring metal conduits for home wiring is one example of this problem. Modern conductor coverings have eliminated the need for metal conduit in numerous applications, but many localities still require it. Its use sells more conduit and labor time for installation.

There are standards nowadays for practically everything, it seems, and consequently we frequently assume that stricter regulation exists than may actually be the case. The public tends to trust implicitly the National Electrical Code in all matters related to power distribution and wiring, but how many people realize that this code, issued by the National Fire Protection Association, is primarily oriented toward *fire* hazards? Only recently has it begun to strengthen provisions against electric shock. Few consumers know that an Underwriter Laboratories seal prominently affixed to the cord of an electrical appliance may pertain to the cord only and not to the rest of the device. In a similar vein, a patent notation inscribed on the handle of a product may refer to just the handle, and then possibly only to the design of the handle's appearance.

Sometimes standards are thought to apply when in actuality there is no standard at all. An example can be found in the widely varying worth and quality of academic degrees that are awarded; doctorates are even available from mail order houses. Appearances can be misleading in this respect. Years ago when competing foreign firms were attempting to corner the South American market for electrical fixtures and appliances, one manufacturing company had a shrewd idea. It equipped its light bulbs with extra long bases and threads. These would fit into the competitors' lamp sockets and its own deep sockets. But the competitors' bulbs would not fit into the deeper sockets of its own fixtures (see sketch in Fig. 3-1). Yet so far as the unsuspecting consumer was concerned, all the light bulbs and sockets continued to look alike.

Problems with the Law in Engineering

The legal regulations which apply to engineering and other professions are becoming more numerous and more specific all the time. We hear many complaints about this trend, and a major effort to "deregulate" various spheres of our lives is currently underway. Nevertheless, we continue to hear cries of "there ought to be a law" whenever a crisis occurs or a special interest is threatened.

This is not surprising. We pride ourselves on being a nation that lives under the rule of law. We even delegate many of our decisions on ethical issues to an interpretation of laws. And yet this emphasis on law can cause problems in regard to ethical conduct quite aside from the more practical issues usually cited by those in favor of deregulation.

For example, one of the greatest moral problems in engineering, and one

FIGURE 3-1
The light bulb story. (a) Long base, deep socket: firm contact; (b) short base, deep socket: no contact; (c) long base, shallow socket: firm contact.

fostered by the very existence of minutely detailed rules, is that of *minimal compliance*. Companies or individuals hunt around for loopholes in the law that will allow them to keep to its letter while violating its spirit. Or hard-pressed engineers sometimes find it convenient to refer to standards with specifications already prepared as a substitute for original thought, perpetuating the "handbook mentality" and the repetition of mistakes.

Minimal compliance led to the tragedy of the *Titanic* (Wade, 68): Why should that ship have been equipped with enough lifeboats to accommodate all its passengers and crew when British regulations in effect at the time did not require it to? Or why should the Tampa Bay Bridge have been designed with possible ship collisions in mind when the code required that only wind loads, not impact loads, be considered in the calculation of horizontal forces?

On the other hand, remedying the situation by continually updating laws or regulations with further specifications may also be counterproductive. Not only will the law inevitably lag behind changes in technology, leading to a judicial vacuum; there is also the danger of overburdening the rules and the regulators. As Robert Kates puts it:

> If cooperation is not forthcoming—if the manufacturer, for example, falsifies or fails to conduct safety tests—there is something akin to the law of infinite regress in which the regulator must intrude more and more expensively into the data collection and evaluation process. In the end, the magnitude of the task overwhelms the regulators (Kates, 32).

The public is frequently lulled into a sense of security by the passage of new laws. Yet many laws are "nonlaws"—that is, laws without enforceable sanctions. These merely serve as window dressing, giving a false display of caring. Or a law may be burdened intentionally by its opponents with so many unreasonable provisions that a repeal will not be far off. Thus there is a need for the critical examination of many laws—and of their sources. Even Adam Smith was moved to make the following observation:

> The proposal of any new law or regulation of commerce which comes from the capitalist class ought always to be listened to with great skepticism, and ought never to be adopted until it has been examined, not only with the most scrupulous, but with the most suspicious attention (Jenkins, 156).

And still another problem with and occasion for frustration over law is the apparent immunity with which powerful interests, including the government, can violate laws when they think they can get away with it by inviting would-be challengers to face them in lengthy and costly court proceedings.

There are many individuals and organizations which see neither moral nor prudential reasons for observing laws. Nor will the law deter white-collar criminals who manage to rationalize their behavior. Self-deceiving rationalizations are easy in the world of business. One glaring example was the 1961 price-fixing scandal in the electric power apparatus industry. Some twenty-nine corporations and forty-five individuals were indicted, including several vice-presidents of General Electric and Westinghouse. Many of the individuals claimed their actions served the public good by stabilizing prices. They knew their acts were illegal, but refused to label themselves as criminals. Crime, they contended, involved harming others, and they did not think their monopoly had hurt anyone (Geis, 122; also see study question 1 below).

The Proper Role of Laws in Engineering

Society's attempts at regulation have indeed often failed, and in various ways. But it would be wrong to write off rule-making and rule-following as futile. Good laws, effectively enforced, clearly produce benefits. They authoritatively establish reasonable minimal standards of professional conduct and provide at least a self-interested motive for most people and corporations to comply with those standards. Moreover, they serve as a powerful support and defense for those who wish to act ethically in situations where ethical conduct might be less than welcome. By being able to point to a law, one can feel freer to act as a responsible engineer.

We contend that to view engineering as social experimentation can provide engineers with a proper perspective on laws and regulations. And the rules which govern engineering practice should not be devised or construed as rules of a game but as rules of responsible experimentation.

Such a view places great responsibility on the engineer, who is intimately connected will his or her "experiment" and responsible for its safe conduct; moreover, it suggests the following conclusions: Precise rules and enforceable sanctions are appropriate in cases of ethical misconduct which involve violations of well-established and regularly-reexamined procedures that have as their purpose the safety and well-being of the public. Little of an experimental nature is probably occurring in such standard activities, and the type of professional conduct required is most likely very clear-cut. In areas where experimentation is involved more substantially, however, rules must not attempt to cover all possible outcomes of an experiment, nor must they force the engineer to adopt a rigidly specified course of action. It is here that regulations should be broad, but so written as to hold the engineer accountable for his or her decisions.

Consider genetic "engineering," for example. One can foresee the time when genetic manipulation will be carried out in a routine manner under strict sets of guidelines. At the present, however, the field is still so new that any rules will invariably leave uncovered some very important safety aspects. Rather than provide unintentional loopholes through such omissions, or to convey a false sense of security to laboratory personnel, it would be better to issue only very general guidelines. The gist of these guidelines would be to place responsibility and accountability for unforeseen consequences on the experimenter.

Study Questions

1 Consult one of the following references about the electric power apparatus price-fixing case of 1961 and discuss the behavior of the engineering managers involved in that scandal in terms of applicable laws, regulations, and codes of ethics: Fuller, Herling, Bane, Geis.

2 Select a regulatory agency and examine its impact on engineering practice. Discuss the roles of engineers, lawyers, and elected representatives in rule-making. If you are not particularly interested in any one agency, the Federal Communications Commission and its problems in allocating broadcasting frequencies may serve as a suitable example.

3 In 1975, Hydrolevel Corporation brought suit against the American Society of Mechanical Engineers (ASME), charging that two ASME volunteers, acting as agents of ASME, had conspired to interpret a section of ASME's Boiler and Pressure Vessel Code in such a manner that Hydrolevel's low-water fuel-cutoff for boilers could not compete with the devices built by the employers of the two volunteers. On May 17, 1982, the Supreme Court upheld the lower courts which had found ASME guilty of violating antitrust provisions and had opened the way for awarding of treble damages. (A U. S. District Court's award of $7.5 million had been found excessive by the Court of Appeals.) Writing on behalf of the 6 to 3 majority, Justice Harry A. Blackmun said: "When ASME's agents act in its name, they are able to affect the lives of large numbers of people and the competitive fortunes of businesses throughout the country. By holding ASME liable under the antitrust laws for the antitrust violations of its agents committed with apparent authority, we recognize the important role of ASME and its agents in the economy, and we help to ensure that standard-setting organizations will act with

care when they permit their agents to speak for them." Acquaint yourself with the particulars of this case and discuss it as an illustration of the possible misuses of standards.

SUMMARY

Engineering is an inherently risky activity, usually conducted with only a partial knowledge of the underlying scientific laws about nature and society and often producing uncertain results and side effects. It lends itself to being viewed as an experiment on a societal scale involving human subjects. While it differs from standard experiments using control groups, it nevertheless imposes the same moral requirements on engineers that are imposed on researchers in other experimental areas involving human subjects. Most important, it requires the following: imaginative forecasting of possible bad side effects, and with this the development of an attitude of "defensive engineering"; careful monitoring of projects; respect for people's rights to give informed consent; and in general that engineers act as *responsible agents*.

Responsible agency, understood as a moral virtue, involves several features: (1) conscientious commitment to live by moral values, (2) a disposition to maintain a comprehensive perspective on the context and possible consequences of one's actions, (3) autonomous, personal, involvement in one's activities, and (4) an acceptance of accountability for the results of one's conduct.

There are many contemporary threats to efforts by engineers to act responsibly, as well as obstacles placed in the way of their respecting the public's right to have the knowledge needed for making informed decisions about engineering products and projects. Those threats and obstacles include the pressures caused by time schedules and organizational rules restricting free speech; the narrow division of labor which tends to cause moral "tunnel vision"; a preoccupation with legalities in a time of proliferating malpractice lawsuits; and the human tendency to divorce oneself from one's actions by placing all responsibility on an "authority" such as one's employer. The three engineers mentioned in connection with the BART case illustrate how to overcome some of these problems, but what happened to them also illustrates the threats to responsible behavior that can be found within large organizations.

Codes of ethics promulgated by professional societies play a variety of roles: (1) inspiration, (2) guidance, (3) support for responsible conduct, (4) deterring and disciplining unethical professional conduct, (5) education and promotion of mutual understanding, (6) contributing to a positive public image of the profession, (7) protecting the status quo and suppressing dissent within the profession, and (8) promoting business interests through restraint of trade.

From the perspective of engineering as social experimentation, the emphasis of codes should be on support of responsible conduct, general guidance, and promotion of mutual understanding rather than on punishment; and the roles of protecting the status quo and promoting business should be avoided altogether. On the other hand, it should be kept in mind that codes are only a small part of engineering ethics. Their brevity renders them overly general and vague, so that some provisions occasionally contradict others. They also represent compromises among many differing viewpoints, the expression and discussion of which must never be stifled. Codes are anything but sacred writ, and should always be viewed as open to critical examination.

A balanced outlook on laws emphasizes both the necessity of laws and regulations and their limitations in governing engineering practice. Laws are necessary because people are not fully responsible and because the competitive nature of our free enterprise system does not always encourage the requisite moral initiative on the part of corporations. Their effects are limited because they encourage minimal compliance with their provisions and tend toward the kind of detailed regulation which can harm productivity and sometimes actually promote violations of the spirit of the law. Moreover, laws inevitably lag behind the pace of technological development.

The model of engineering as social experimentation allows for the importance of clear laws, effectively enforced. But it places equal emphasis on the moral responsibility of engineers—an emphasis that goes beyond merely following laws and is especially vital for those working at the frontiers of technological development.

CHAPTER 4

THE ENGINEER'S CONCERN FOR SAFETY

Pilot Dan Gellert was flying an Eastern Airlines Lockheed L-1011, cruising at an altitude of 10,000 feet, when he inadvertently dropped his flight plan. Being on autopilot control, he casually leaned down to pick it up. In doing so, he bumped the control stick. This should not have mattered, but immediately the plane went into a steep dive, scaring the 230 passengers no end. Badly shaken himself, Gellert was nevertheless able to grab the control stick and ease the plane back onto course. Though much altitude had been lost, the altimeter still read a stable 10,000 feet.

Not long before this incident, one of Gellert's colleagues had been in a flight trainer when the autopilot and the flight trainer disengaged, producing a crash on an automatic landing approach. Fortunately it all happened on simulation. But just a short time later, an Eastern Airlines L-1011 actually crashed on approach to Miami. On that flight there seemed to have been some problem with the landing gear, so the plane had been placed on autopilot at 2000 feet while the crew investigated the trouble. Four minutes later, after apparently losing altitude gradually and evidently without warning, it crashed in the Everglades, killing 103 people.

A year later Gellert was again flying an L-1011 and the autopilot disengaged once more when it should not have done so. The plane was supposedly at 500 feet and on the proper glide slope to landing as it broke through a cloud cover. Suddenly realizing it was only at 200 feet and above a densely populated area, however, the crew had to engage the plane's full takeoff power to make the runway safely.

The L-1011 incidents point out how vulnerable our intricate machines and control systems can be, how failures in their working can be caused by unanticipated circumstances, and how important it is to design for proper human-machine interactions whenever human safety is involved. The reader who is curious to find out what happened to Gellert's frustrating efforts to seek corrective action is referred to his account, "Insisting on Safety in the Skies" (Westin, 1981, 17–30). Here we turn to a more general discussion of the role of safety as seen by the public and the engineer.

For the purposes of our discussion, the term "public" will mean lay people who use a product of technology without an intimate knowledge of its design, manufacture, and operating characteristics. Thus airplane passengers constitute the flying "public," even though some of them might be engineers well informed about other fields of technology. The "engineer," in contrast, is a person in possession of special knowledge about the product by virtue of having participated in its design, its manufacture, or the application decisions made in regard to it, whether as a member of the organization selling it or that purchasing it. The pilot of an airplane fits in between the two categories—Gellert was given a plane to command, but he had no control over it in terms of design or modifications.

Thus there are typically several groups of people involved in safety matters, each with its own interests at stake. If we now consider that within each group there are differences of opinion regarding what is safe and what is not, it becomes obvious that "safety" can be an elusive term. It behooves us, therefore, to decide upon a working definition of safety. And to help us understand the subject even more fully, we will discuss it in conjunction with the term "risk." Following a look at these basic concepts, we will then turn to safety and risk assessment and methods of reducing risk (increasing safety). Finally, by way of a case study of the Three Mile Island nuclear reactor incident, we will consider the legal, political, and ethical implications of safety—in short, why safety should be of paramount concern to the engineer.

SAFETY AND RISK

We demand safe products and services because we do not wish to be threatened by potential harm, but we also realize that we may have to pay for this safety. To complicate matters, what may be safe enough for one person may not be so for someone else—either because of different perceptions about what is safe or because of different predispositions to harm. For example, a power saw in the hands of a child will never be as safe as it can be in the hands of an adult. And a sick adult is more prone to suffer ill effects from air pollution than is a healthy adult.

Absolute safety, in the sense of a degree of safety which satisfies all individuals or groups under all conditions, is neither attainable nor affordable. Yet it is important for our discussion here that we come to some understanding of what *we* mean by safety.

The Concept of Safety

One approach to defining safety would be to render the notion thoroughly subjective by defining it in terms of whatever risks a person judges to be acceptable. Such a definition was given by William W. Lowrance: *"A thing is safe if its risks are judged to be acceptable"* (Lowrance, 8). This approach helps underscore the notion that judgments about safety are tacitly value judgments about what is acceptable in the way of risk to a given person or group. Differences in appraisals of safety are thus correctly seen as reflecting differences in values.

Lowrance's definition, however, needs to be modified, for it departs too far from our common understanding of safety. This can be shown by three considerations. Imagine, first, a case where we seriously *underestimate* the risks of something—say of using a toaster we see at a garage sale. On the basis of that mistaken view, we judge it to be very safe and buy it. On taking it home and trying to make toast with it, however, it sends us to the hospital with a severe electric shock and burn. Using the ordinary notion of safety, we conclude we were wrong in our earlier judgment: The toaster was not safe at all! Given our values and our needs, its risks should not have been judged acceptable earlier. Yet by Lowrance's definition we would be forced to say that prior to the accident the toaster was entirely safe since, after all, at that time we had judged the risks to be acceptable.

Consider, second, the case where we grossly *overestimate* the risks of something. For example, we irrationally think fluoride in drinking water will kill a third of the populace. According to Lowrance's definition, the fluoridated water is unsafe, since we judge its risks to be unacceptable. It would, moreover, be impossible for someone to reason with us to prove that the water is in reality safe. For again, according to his definition, the water became unsafe the moment we judged the risks of using it to be unacceptable for us. But of course, our ordinary concept of safety allows us to say the water has been perfectly safe all along, in spite of such irrational judgments.

Third, there is the situation where a group makes no judgment at all about whether the risks of a thing are acceptable or not—they simply do not think about it. By Lowrance's definition this means the thing is neither safe nor unsafe with respect to that group. Yet this is somewhat paradoxical, given our ordinary ways of thinking about safety. For example, we normally say that some cars are safe and others unsafe, even though many people may never even think about the safety of the cars they drive.

The point is that there must be at least some objective point of reference outside ourselves which allows us to decide whether our judgments about safety are correct or not. An adequate definition should capture this element, without omitting the insight already noted that safety judgments are relative to people's value perspectives (Council for Science and Society, 13).

We propose to adopt as our working definition a modified version of Lowrance's definition:

A thing is safe if, were its risks fully known, those risks would be judged acceptable in light of settled value principles.

More fully,

A thing is safe (to a certain degree) with respect to a given person or group at a given time if, were they fully aware of its risks and expressing their most settled values, they would judge those risks to be acceptable (to a certain degree).

The objections to Lowrance's definition raised by the examples given above are met by the new definition's "knowledge" condition. And the further condition that a judgment about safety express "settled value principles" helps to rule out as irrelevant many other types of judgments that could be problematic: For example, judgments made while heavily intoxicated would not count.

Thus in our view safety is a matter of how people *would* find risks acceptable or unacceptable *if* they knew the risks and were basing their judgments on their most settled value perspectives. To this extent safety is an *objective* matter. It is a *subjective* matter to the extent that value perspectives differ. In what follows we will usually speak of safety simply as acceptable risk. But this is merely for convenience, and should be interpreted as an endorsement of Lowrance's definition only as we have qualified it.

Safety is frequently thought of in terms of degrees and comparisons. We speak of something as "fairly safe" or "relatively safe" (compared with similar things). Using our definition, this translates as the degree to which a person or group, judging on the basis of their settled values, would decide that the risks of something are more or less acceptable in comparison with the risks of some other thing. For example, when we say that airplane travel is safer than automobile travel, we mean that for each mile traveled it leads to fewer deaths and injuries—the risky elements which our settled values lead us to avoid.

We interpret "things" to include products as well as services, institutional processes, and disaster protection. The definition could therefore extend into medicine, finance, and international affairs, to mention just a few of the "things" and "services" organized by people. And for engineers the definition would extend into the safe operation of systems and the prevention of natural or people-caused disasters.

Risks

We say a thing is "not safe" if it exposes us to unacceptable risks. What is meant by "risk?" *A risk is the potential that something unwanted and harmful may occur.* Rowe refers to the "potential for the realization of unwanted consequences from impending events" (Rowe, 1977, 24). Thus a future, possible, occurrence of harm is postulated.

Risk, like harm, is a wide concept covering many different types of unwanted occurrences. In regard to technology, it can equally well include dangers of bodily harm, of economic loss, or of environmental degradation. These in turn can be caused by delayed job completion, faulty products or systems, and economically or environmentally injurious solutions to technological problems.

Good engineering practice has always been concerned with safety. But as technology's influence on society has grown, so has public concern about technological risks increased. In addition to measurable and identifiable hazards arising from the use of consumer products and from production processes in factories, some of the less obvious effects of technology are now also making their way to public consciousness. And while the latter are often referred to as "new risks," many of them have existed for some time. They are new only in the sense that (1) they are now identifiable (because of changes in magnitude of the risks they present, having passed a certain threshold of accumulation in our environment, or because of a change in measuring techniques, allowing detection of hitherto unnoticeable traces), or (2) the public's perception of them has changed (because of education, experience, or media attention, or because of a reduction in other hitherto dominant and masking risks).

Meanwhile, natural hazards continue to threaten human populations. Technology has greatly reduced the scope of some of those, such as floods, but at the same time it has increased our vulnerability to others as they affect our ever greater concentrations of population and cause greater damage to our finely tuned, technological networks.

A word here should be said about disasters. A disaster does not take place until a seriously disruptive event coincides with a state of insufficient preparedness (Dynes). Hence the *Titanic's* collision with an iceberg did not in itself constitute a disaster, but rather an emergency. The real disaster in terms of lives lost came about because emergency preparedness was inadequate: There were too few lifeboats, and there had been no lifeboat drills worth mentioning.

And if a disaster emerges from a combination of factors, so too does a risk—in the latter case from a combination of probability and consequence. The probabilistic aspects arise out of uncertainties over the event and who its victims will be, and the severity of the risk is judged by its nature and possible consequences (Rowe, 1977, 28). All this, of course, is related to the notion of experiment, for we are speaking of the "experimental" risks connected with the introduction of new technology, the risks associated with new applications of familiar technology, and the risks arising from attempts at disaster control. So again we shall find our paradigm of engineering as social experimentation to be of some use as we unravel the ethical implications of safety and risk in regard to engineered products.

Acceptability of Risk

Having adopted a modified version of Lowrance's definition of safety as acceptable risk, we need to examine the idea of acceptability more closely. William D. Rowe says that "a risk is acceptable when those affected are generally no longer (or not) apprehensive about it" (Rowe, 1979, 328). Apprehensiveness depends to a large extent on how the risk is perceived. This is influenced by such factors as whether or not the risk is assumed voluntarily; the effects of knowledge on how the probabilities of harm (or benefit) are perceived; job-related or other pressures that cause people to be aware of or (alternatively) to overlook risks; whether or not the effects of a risky activity or situation are immediately noticeable or are close at hand; and whether or not the potential victims are identifiable beforehand. Let us illustrate these elements of risk perception by means of some examples.

Voluntarism and Control   John and Ann Smith and their children enjoy riding motorcycles in rough terrain for amusement. They take *voluntary* risks—that is part of being engaged in such a potentially dangerous sport. They do not expect the manufacturer of their dirt-bikes to adhere to the same standards of safety as they would the makers of a passenger car used for daily commuting. The bikes should be sturdy, but guards covering exposed parts of the engine, padded instrument panels, collapsible steering mechanisms, or emergency brakes are clearly unnecessary, if not inappropriate.

John and Ann also live near a chemical plant. It is the only area in which they can afford to live, and it is near the shipyard where they both work. At home they suffer from some air pollution, and there are some toxic wastes in the ground. Official inspectors tell them not to worry. Nevertheless they do, and they think they have reason to complain—they do not care to be exposed to risks from a chemical plant with which they have no relationship except on an involuntary basis. Any beneficial link to the plant through consumer products or other possible connections is very remote and, moreover, subject to choice.

John and Ann behave as most of us would under the circumstances: We are much less apprehensive about the risks to which we expose ourselves voluntarily than those to which we are exposed involuntarily. "We are loath to let others do unto us what we happily do to ourselves" (Starr, quoted in Lowrance, 87). In terms of our "engineering as social experimentation"paradigm, people are more willing to be the subjects of their own experiments (social or not) than of someone else's.

Intimately connected with this notion of voluntarism is the matter of *control*. The Smiths choose where and when they will ride their bikes. They have selected their machines and they are proud of how well they can control them (or think they can). They are aware of accident figures, but they tell themselves those apply to other riders, not to them. In this manner they may well display the characteristically unrealistic confidence of most people when they believe hazards to be under their control (Slovic, Fischhoff, and Lichtenstein, April 1979, June 1980, and May 1979). But still, riding motorbikes cross-country, skiing, hanggliding, horseback riding, boxing, and other hazardous sports are usually carried out under the implied control of the participants, which is a good part of why they are engaged in voluntarily at all and why their enthusiasts worry less about their risks than the dangers of, say, air pollution or airline safety. Another reason for not worrying so much about the consequences of these sports is that rarely does any one accident injure any appreciable number of innocent bystanders.

Effect of Information on Risk Assessments   The manner in which information necessary for decision making is presented can greatly influence how risks are perceived. The Smiths are careless about using seat belts in their car. They know that the probability of their having an accident on any one trip is infinitesimally small. Had they been told, however, that in the course of 50 years of driving, at 800 trips per year, there is a probability of 1 in 3 that they will receive at least one disabling injury, their seat belt habits (and their attitude about seat belt laws) would likely be different (Arnould, 35).

Studies have verified that a change in the manner by which information about a danger is presented can lead to a striking reversal of preferences about how to deal with that danger. Consider, for example, an experiment in which two groups of around 150 people each were told about the strategies available for combatting a disease. The first group was given the following description:

Imagine that the U.S. is preparing for the outbreak of an unusual Asian disease, which is expected to kill 600 people. Two alternative programs to combat the disease have been proposed. Assume that the exact scientific estimate of the consequences of the programs are as follows:
   If Program A is adopted, 200 people will be saved. . . .

If Program B is adopted, there is ⅓ probability that 600 people will be saved, and ⅔ probability that no people will be saved. . . .
Which of the two programs would you favor? (Tversky and Kahneman, 453)

72 percent of the respondents selected Program A, and only 28 percent selected Program B. Evidently the vivid prospect of saving 200 people led many of them to feel averse to taking a risk on possibly saving all 600 lives.

The second group was given the same problem and the same two options, but the options were worded differently:

If Program C is adopted 400 people will die. . . .
If Program D is adopted there is ⅓ probability that nobody will die, and ⅔ probability that 600 people will die. . . .
Which of the two programs would you favor? (Tversky and Kahneman, 453)

This time only 22 percent chose Program C, which is the same as Program A. 78 percent chose Program D, which is identical to Program B.

One conclusion the experiment provides is that options perceived as yielding firm gains will tend to be preferred over those whose gains are perceived as risky or only probable. A second conclusion is that options emphasizing firm losses will tend to be avoided in favor of those whose chances of success are perceived as probable. In short, people tend to be more willing to take risks in order to avoid perceived firm losses than they are to win only possible gains.

The difference in perception of probable gain and probable loss is illustrated graphically in Fig. 4–1. The typical risk-benefit value function shown there drops more steeply on the loss portion than it rises on the gain portion. We have included on this graph a loss side threshold, as does Rowe (Rowe, 1979, 331). The threshold is ascribable to the human habit of ignoring smaller hazards in order to avoid anxiety overload and means that no value is attached to a first small amount of loss or that no effort is expended to overcome the loss. We have added a similar, though smaller, threshold on the gain side to account for the normal human inertia and a certain amount of inherent generosity that often restrain people in how they set about seeking their own gain.

The thresholds are significant because they remind us that different people have different tolerances for specific conditions. Someone with a respiratory illness, for example, will react to the smallest amount of air pollution; thus the threshold for pollution is near zero for that person. An entire population living near an oil refinery, however, will have a fairly high tolerance (i.e., a large threshold) as far as automobile emissions alone are concerned.

Job-Related Risks John Smith's work in the shipyard has in the past exposed him to asbestos. He is aware now of the high percentage of

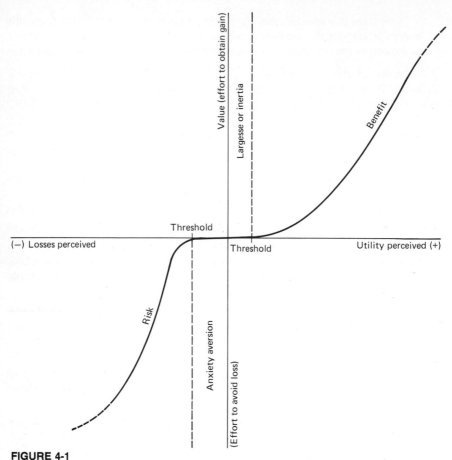

FIGURE 4-1
Typical risk-benefit value function.

asbestosis cases among his coworkers, and after consulting his own physician was told that he was slightly affected himself. Even Ann, who works in a clerical position at the shipyard, has shown symptoms of asbestosis as a result of handling her husband's clothes. Earlier John saw no point to "all the fuss stirred up by some do-gooders." He figured that he was being paid to do a job, he felt the masks which were handed out occasionally gave him sufficient protection, and he thought the company physician was giving him a clean bill of health.

In this regard John's thinking is similar to that of many workers who take risks on their jobs in stride, and sometimes even approach them with a bravado attitude. Of course exposure to risks on a job is in a sense voluntary since one can always refuse to submit oneself to them, and workers perhaps even have some control over how their work is carried out. But often employees have little choice other than to stick with what is for them the only available job and to do as they are told on it. Unions and occupational health and safety regulations can correct the worst situations, but standards regulating conditions in the workplace (its air quality, for instance) are generally still far below those which regulate conditions in our general (public) environment. It may be argued that the "public" encompasses many people of only marginal health whose low thresholds for pollution demand a fairly clean environment. On the other hand factory workers are seldom carefully screened for their work. And in all but the most severe environments (those conducive to black lung or brown lung, for instance), unions display little desire for change lest necessary modifications of the workplace force employers out of business altogether.

Engineers who design and equip work stations must take into account the cavalier attitude toward safety shown by many workers, especially when their pay is on a piecework basis. And when one worker complains about unsafe conditions, but others do not, the complaint should not be dismissed as coming from a crackpot. Any report regarding unsafe conditions merits serious attention.

Pilot Dan Gellert's reports about problems with the L-1011s should have been looked into promptly. Later attempts to discredit him through psychiatric examinations may have been used partly as a legal ploy, but management might also have found it hard to believe that anyone could be so particular about safety. And yet Gellert was properly concerned about his responsibilities as a pilot—and all the more so because he had an airliner full of passengers and crew relying on his skill and the integrity of the plane rather than just himself to take into consideration. Which brings us to yet another factor that colors our perception of risks: their magnitude and proximity.

Magnitude and Proximity Our reaction to risk is affected by the *dread* of a possible mishap, both in terms of its magnitude and of the personal identification or relationship we may have with the potential victims. A single major airplane crash, the specter of a child we know trapped in a cave-in—these affect us more acutely than the ongoing but anonymous carnage on the highways, at least until someone close to us is killed in a car accident.

In terms of numbers alone we feel much more keenly about a potential risk if one of us out of a group of 20 intimate friends is likely to be subjected to great harm than if it might affect, say, 50 strangers out of a larger group of 1000. This proximity effect is noticeable in the time domain as well. A future risk is easily dismissed by various rationalizations including (1) the attitude of "out of sight, out of mind," (2) the assumption that predictions for the future must be discounted by using lower probabilities, or (3) the belief that a countermeasure will be found in time.

Lessons for the Engineer

Engineers in their work face two problems in regard to public conceptions of safety. On the one hand there is the overoptimistic attitude that things which are familiar, which have not hurt us before, and over which we have some control, present no real risks. On the other hand is the dread people feel when accidents kill or maim in large numbers or harm those we know, even though statistically speaking such accidents might occur infrequently.

Leaders of industry are sometimes heard to proclaim that those who fear the effects of air pollution, toxic wastes, or nuclear power are emotional and irrational or politically motivated. This in our view is a misperception of legitimate concerns expressed publicly by thoughtful citizens. It is important that engineers recognize as part of their work reality such widely held perceptions of risk and take them into account in their designs. It is not wise to proceed under the assumption that "education" will quickly change the public's underestimation or overestimation of risk. As Paul Slovic, Baruch Fischhoff, and Sarah Lichtenstein point out:

> Another barrier to educational attempts is that people's beliefs change slowly and are extraordinarily resistant to new information. Research in social psychology has often demonstrated that once formed, people's initial impressions tend to structure the way they interpret subsequent information. They give full weight to evidence that is consistent with their initial beliefs while dismissing contrary evidence as unreliable, erroneous, or unrepresentative. Whereas opponents of nuclear power believe the accident at Three Mile Island "proved" how dangerous reactors are, proponents felt that it confirmed their faith in the effectiveness of the multiple safety and containment system (Slovic, Fischhoff, and Lichtenstein, 1980, 48).

And in regard to professionals they continue:

> Since even well-informed citizens have difficulty in judging risk accurately, and the cognitive functioning of experts appears to be basically like that of everyone else, it seems clear that no one person or profession knows how to get the right answers. The best we can hope to do is to keep the particular kinds of mistakes to which each of us is prone to a minimum by being more aware of our tendency to make mistakes (Slovic, Fischhoff, and Lichtenstein, 1980, 48).

Study Questions

1 Describe a real or imagined traffic problem in your neighborhood involving children and elderly people who find it difficult to cross a busy street. Put yourself in the position of (a) a commuter traveling to work on that street, (b) the parent of a

child, or the relative of an older person, who has to cross that street on occasion, (c) a police officer assigned to keep the traffic moving on that street, and (d) the town's traffic engineer working under a tight budget.

Describe how in these various roles you might react to (i) complaints about conditions dangerous to pedestrians at that crossing and (ii) requests for a pedestrian crossing protected by traffic lights.

2 The Chinese ideograph for "crisis" consists of the characters for (a) "danger" and (b) "chance" or "opportunity" (also used for "machine"). What insight does this combination offer in regard to the nature of a crisis, whether natural or caused by people?

(a)   (b)

3 In some technologically advanced nations, a number of industries which have found themselves restricted by safety regulations have resorted to dumping their products on, or moving their production processes to, less developed countries where higher risks are tolerated. Examples are the dumping of unsafe or ineffective drugs on the Third World by pharmaceutical companies from Western Europe, Communist Bloc countries, Japan, and the United States (Silverman, Lee, and Lydecker) and the transfer of asbestos processing from the United States to Mexico (Shue, 586). To what extent do differences in perception of risk justify the transfer of such merchandise and production processes to other countries? Is this an activity that can or should be regulated?

4 The industrial accident described below and illustrated in Fig. 4–2 raises several issues of ethical import. Identify and discuss them.

The following news story is based on the Nassau edition of Newsday, the Long Island, N.Y. newspaper, April 24, 1981.

Inadequate safety precautions and an accident inside an empty water tank caused the deaths of two workmen in New Jersey on April 23. At 4 p.m., a scaffold inside the tank collapsed and caused the two men painting the tank to fall to the bottom. Stranded there, they were overcome by paint fumes and eventually lost consciousness. John Bakalopoulos, 34, of Brooklyn, N.Y. and Leslie Salomon, 31, also of Brooklyn, were not wearing oxygen masks. The Suffolk County Water Authority's contract for the painting job specified that workmen wear "air hoods," masks connected to air compressors. The masks were available, but Bakalopoulos and Salomon had decided not to wear them because they were unwieldy. Instead, Bakalopoulos wore a thin gauze mask designed to filter out dust and paint particles. Salomon wore no mask.

Peter Koustas, the safety man who was handling the compressor and paint feed outside the tank, asked a nearby resident to call firemen as soon as he realized the scaffold had collapsed. Then he rushed into the tank with no oxygen mask, and he, too, was overcome by the fumes and lost consciousness.

The men lay unconscious for hours as rescue efforts of more than 100 policemen, firemen, and volunteers were hampered by bad weather. Intense fog,

FIGURE 4-2
Accident in a water tank *(Source: Opflow, American Water Works Association, vol. 7, no. 6, June 1981, p. 3, using material from Newsday, Nassau edition, Long Island, NY, April 24, 1981.)*

rain, and high winds made climbing the tank difficult and restricted the use of machinery. Several men collapsed from fatigue.

Inside the tank, conditions were worse. Because of the heavy fumes, rescuers used only hand-held, battery-powered lights, fearing that sparks from electric lights might cause an explosion. Lt. Larry Viverito, 38, a Centereach, N.Y. volunteer fireman, was overcome by fumes 65 ft (20 m) above the floor of the tank. Fellow rescuers had to pull him out.

Rescuer John Flynn, a veteran mountain climber, said he hoped he would never have to go through anything like that night again. For five hours he set up block-and-tackle pulleys, tied knots, adjusted straps on stretchers, and attached safety lines and double safety lines. The interior of the tank was as blindingly white as an Alpine blizzard—completely and nauseatingly disorienting. Fans that had been set up to pull fresh air into the tank caused deafening noise.

When Flynn first reached the tank floor, he stepped into the wet paint and began to slide toward the uncovered 4-ft (1.2-m) opening to the feeder pipe in the center of the floor. Flynn was able to stop sliding, but John Bakalopoulos wasn't as fortunate.

As rescuers watched helplessly, Bakalopoulos, still out of reach, stirred, rolled over, and in the slippery paint slid into the feeder pipe. He plunged 110 ft (34 m) to the bottom.

Bakalopoulos was dead on arrival at the University Hospital in Stony Brook, N.Y. Peter Koustas, rescued at 1:45 a.m. and suffering from hypothermia, died the following morning when his heart failed and he could not be revived. Only Leslie Salomon survived. *(Quoted with permission from Opflow, Am. Water Works Assoc., Vol. 7, No. 6, June 1981, p. 3, and from Newsday.)*

ASSESSMENT OF SAFETY AND RISK

Absolute safety is not attainable, and any improvement in safety as it relates to an engineered product is often accompanied by an increase in cost of that product. On the other hand, products that are not safe incur secondary costs to the manufacturer beyond the primary (production) costs that must also be taken into account—costs associated with warranty expenses, loss of customer goodwill and even loss of customers due to injuries sustained from use of the product, possible downtime in the manufacturing process, and so forth. It is therefore important for manufacturers and users alike to reach some understanding of the risks connected with any given product and of what it might cost to reduce those risks or not reduce them.

As Fig. 4–3 indicates, an emphasis on high safety (low risks) incurs high primary costs, but secondary costs are low. At the other extreme of high risks (low safety), one saves on primary costs but pays dearly because of high secondary costs. In between, where the slopes of the primary and secondary cost curves are equal, is the point of minimum total cost. If all costs were quantifiable, that optimum point would be the goal to reach for. But before we crank up our computers to home in on such an optimal

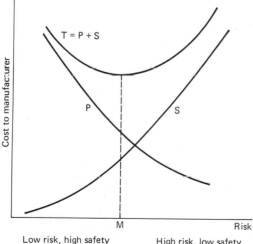

FIGURE 4-3
Why both low-risk and high-risk products are costly. P = primary cost of product, including cost of safety measures involved; S = secondary costs, including warranties, loss of customer goodwill, litigation costs, costs of downtime, etc.; T = total cost. Minimum total cost occurs at M, where incremental savings in primary cost (slope of P) are offset by an equal incremental increase in secondary cost (slope of S).

design, we must be clear about how to determine risk (to be discussed in this section) and how to compare losses with benefits (to be covered in the next section).

Knowledge of Risk

One would think that experience and historical data would provide good information about the safety of standard products. While much has been collected and published, however, there are gaps because (1) there are some industries where information is not freely shared, and (2) there are always new applications of old technology which render the available information less useful.

Engineers are by nature inclined to share information freely. It is in this spirit, according to R.R. Whyte in *Engineering Progress through Trouble*, that Robert Stephenson, famous bridge builder during the first half of the nineteenth century, took the following position upon reviewing a technical paper:

> . . . he hoped that all the casualties and accidents, which had occurred during their progress, would be noticed in revising the paper; for nothing was so instructive to the younger Members of the Profession, as records of accidents in large works, and of the means employed in repairing the damage. A faithful account of those accidents, and of the means by which the consequences were met, was really more valuable than a description of the most successful works. The older Engineers derived their most useful store of experience from the observations of those casualties which had occurred to their own and to other works, and it was most important that they should be faithfully recorded in the archives of the Institution (R. R. Whyte, v).

We also take the following account from pp. 54–57 of the same book: In 1950 the chief engineer for a British manufacturer of large generators invited his competitors to study the failure of a rotor endbell during overspeed tests. The endbell is a retaining sleeve which holds in place the endturns of the rotor winding so they will not fly apart at the high speed of the turbine generator. Because the endbell has to be made of nonmagnetic steel, it presents severe metallurgical problems. This engineer's action led to a similar frank divulgence of information on the occasion of another, similar, failure in Canada some years later.

Such examples are noteworthy because they are so rare—which is regrettable. Too many companies believe that releasing technical information might hurt their competitive position, if not place them at a disadvantage in case of litigation. So it is that new engineers and new companies usually have to learn from scratch, although sometimes past experience is used effectively to educate beginners (see, for example, the textbook on soil mechanics and foundations, *Introductory Soil Mechanics and Foundations*, by George B. Sowers and George F. Sowers).

Uncertainties in Design

Risk is seldom intentionally designed into a product. It arises because of the many uncertainties faced by the design engineer, the manufacturing engineer, and even the sales and applications engineer.

To start with, there is the purpose of a design. Let us consider an airliner. Is it meant to maximize profits for the airline, or is it intended to give the highest possible return on investment? The answer to that question is important to the company because on it hinge different decisions and their outcomes and the possibility of the airline's economic success or ruin. Investing $50 million on a jumbo jet to bring in maximum profits of, say, $10 million during a given time period involves a lower return-on-investment than spending $24 million on a medium-sized jet to bring in a return of $6 million in that same period.

Apart from uncertainties about applications of a product, there are uncertainties regarding the materials of which it is made and the quality of skill that goes into designing and manufacturing it. For example, changing economic realities or hitherto unfamiliar environmental conditions such as extremely low temperatures may affect how a product is to be designed. A typical "handbook engineer" who extrapolates tabulated values without regard to their implied limits under different conditions will not fare well under such circumstances. Even a careful analyst will face difficulties when confronted with data such as those illustrated in Fig. 4–4, which gives the thermal conductivity of copper over wide ranges of temperature as observed by different investigators.

In other situations, designs which do quite well under static loads fail under dynamic loading. A famous example is the wooden bridge that collapsed when a contingent of Napoleon's army crossed it marching in step. This even happened to one of Robert Stephenson's steel bridges, which shook violently under a contingent of marching British troops. Ever since then, soldiers are under orders to fall out of step when crossing a bridge. Wind can also cause severe vibrations: The Tacoma Narrows Bridge, which collapsed some years ago, and a high tension power line across the Bosphorus that broke after a short-circuit caused by swinging cables are but two of such examples.

Even when the properties of materials under standard operating conditions are well known, caution is called for in using them. Only recently a bridge, which had just replaced an old and trusted ferry service, was found to have been built from steel so unsafe that the entire structure had to be

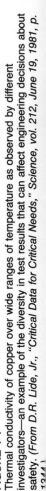

FIGURE 4-4

Thermal conductivity of copper over wide ranges of temperature as observed by different investigators—an example of the diversity in test results that can affect engineering decisions about safety. (*From D.R. Lide, Jr., "Critical Data for Critical Needs," Science, vol. 212, June 19, 1981, p. 1244.*)

closed. In the meantime the ferries had disappeared! While strength tests are usually carried out on concrete, the strength of steel is often taken too much for granted.

Such drastic variations from the standard quality of a given grade of steel are rather exceptional. More typically the variations are small. Nevertheless the design engineer should realize that the supplier's data on items like steel, resistors, insulation, optical glass, and so forth apply to statistical averages only. Individual components can vary considerably from the mean.

Engineers traditionally have coped with such uncertainties about materials or components, as well as incomplete knowledge about the actual operating conditions of their product, by introducing a comfortable "factor of safety." That factor is intended to protect against problems arising when stresses due to anticipated loads (duty) and stresses the product as designed is supposed to withstand (strength or capability) depart from their expected values. Stresses can be of a mechanical or any other nature, for example, an electric field gradient to which an insulator is exposed, or the traffic density at an intersection.

A product may be said to be safe if its capability exceeds its duty. But this presupposes exact knowledge of actual capability and actual duty. In reality the stress calculated by the engineer for a given condition of loading and the stress which ultimately materializes at the loading may vary quite a bit. This is because each component in an assembly has been allowed certain tolerances in its physical dimensions and properties—otherwise the production cost would be prohibitive. The result is that the assembly's capability as a whole cannot be given by a single numerical value but must be expressed as a probability density which can be graphically depicted as a "capability" curve (see Fig. 4–5a and b). For a given point on a capability curve, the value along the vertical axis gives the probability that the capability, or strength, is equal to the corresponding value along the horizontal axis.

A similar curve can be constructed for the duty which the assembly will actually experience. The stress exposure varies because of differences in loads, environmental conditions, or the manner in which the product is used. Associated with the capability and duty curves are nominal or, statistically speaking, expected values C and D. We often think and act only in terms of nominal or expected values. And with such a deterministic frame of mind, we may find it difficult to conceive of engineering as involving experimentation. The "safety factor" C/D rests comfortably with our consciences. But how sure can we be that our materials are truly close to their specified nominal properties? Or that the loads will not vary too widely from their anticipated values or occur in environments hostile to the proper functioning of the materials? It is entirely conceivable that the capability and duty curves will assume flatter shapes because of increased variances as

FIGURE 4-5
Probability density curves for stress in an engineered system. (a) Variability of stresses in a relatively safe case; (b) an unsafe case due to overlap in duty and capability stresses.

shown in Fig. 4-5b, than they would have under normal conditions, as in Fig. 4–5a. And Fig. 4–5b shows how it is probable that load stress may exceed design stress. A measure of the probability that this might happen and failure occur is given by the size of the shaded area. Mathematical treatment of this topic is offered by, among others, Edward B. Haugen, who reminds his readers how "the safety factor concept completely ignores the facts of variability that result in different reliabilities for the same safety factor" (Haugen, 5).

A more appropriate measure of safety would be the "margin of safety," which is shown in Fig. 4–5a. If it is difficult to compute such a margin of safety for ordinary loads used every day, imagine the added difficulties that arise when repeatedly changing loads have to be considered. As F. Nixon, N. E. Frost, and K. J. March point out:

> The use of a general safety margin on the static strength of a material (or structure) is quite unsuitable for cyclic loadings. For example, local static stresses are usually unimportant in ductile materials because regions which become over-stressed may yield, with the result that the stress becomes more uniformly distributed. However, when the loading is cyclic, yield can affect only the mean value of the local stresses to which any local discontinuities give rise. Thus, the merit of a structure subjected to dynamic loadings depends to a great extent on the skill of the designer in avoiding unnecessary concentrations of stress (Nixon, Frost, and March, 137).

Testing for Safety

The widely proclaimed "safety factor" thus obviously has some serious conceptual flaws. So what can the engineer do to assure safety? "Rely on experience" was mentioned at the outset of this discussion. But it was also pointed out that experience gained by one engineer is all too often not passed on to others, especially if the experience was professionally embarrassing. Bad news travels fast, we agree, but usually it travels unaccompanied by hard facts.

Another way of gaining experience is through tests. Under certain conditions this can be a valuable source of information, especially if the testing of materials or a product is carried out to destruction. An example of such testing was that performed on a Comet aircraft fuselage after two of those early jets had crashed in service in 1954. One conclusion emerged from initial investigations:

> . . . fatigue of the pressure cabin. The only possible objection was that fatigue so early in the lives of the two aircraft seemed at that time incredible. These arguments provided the justification and the objective for the experimental attack that then followed.
>
> The fatigue-testing of a complete full-size pressure cabin and fuselage went beyond anything ever before attempted. For safety, water was used as the testing medium, and the whole fuselage was immersed in a 250,000 gal. tank.
>
> It was clearly necessary to introduce every conceivable adverse factor in addition to pressure. In particular, the wings had to be included in the test and given the appropriate load fluctuations. Total immersion of the wings was considered, but rejected in favor of their projecting through the sides of the tank, the water being held back by flexible sleeves that allowed the wings to articulate.
>
> With [special apparatus] the cabin was made to breathe in and out as the

wings were forced to bend up and down realistically. The test went on night and day until a piece of pressure cabin 20 ft² in area was pushed out almost instantaneously. The failure was initiated by a fatigue crack at a corner of one of the forward windows. In the air, of course, this would have meant an explosion. (De Havilland and Walker, 53).

The reader should note that this test was carried out *after* real accidents had occurred. The more usual procedure is to subject prototypes to testing. Yet there are severe limitations to relying on prototype tests, as R. R. Whyte points out:

> . . . successful prototype testing and prolonged test-bed running also do not ensure a reliable machine. . . . Like the designer, the development engineer is seldom allowed sufficient time. In many cases all he has time for is a functional test too short to reveal potential weaknesses in the design such as a fatal resonant frequency, for example. In others all that is stipulated even by customers who should know better, is that a single prototype should satisfactorily pass a "type-approval test"; maybe a few hundred operations or hours of running.
>
> Such a test can do no more than demonstrate that one individual product— seldom representative of large quantities—is capable of passing the stipulated test. It can do little to demonstrate the ultimate capability of the design. It can do nothing to indicate the variability in capability which is likely to exist when numbers of similar products are to be produced or the long-term effects such as corrosion or fatigue (R. R. Whyte, 139).

Even prototype tests and routine quality assurance tests are frequently not carried out properly. Suppliers of rifles and ammunition to the Armed Forces have been found to have committed fraud in the testing of their products; the Alaska pipeline was plagued with poor welds and inadequate testing; the Ford Motor Co. at one time was found to have falsified emission test data; the B. F. Goodrich Co. delivered an aircraft brake for test flights on a new Air Force plane although the brake failed to meet Air Force specifications even according to a common sense interpretation of those specifications.

In short, we cannot trust testing procedures uncritically. Time pressure, as we have said, is one factor contributing to shoddy testing. The boredom of routine that tempts one to quickly duplicate test data of a repetitive nature can be another. At times there is pressure from management to "fudge the data for now" since "by the time we get into production we will have ironed out the problem." And not to be overlooked is the problem of outright fraud, as when testers are bribed to pass faulty items or when no tester was on the job although testing is claimed to have been undertaken. Conscientious engineers had therefore better make occasional spot checks on their own unless the organization they work for operates an independent testing service free of production pressures.

When Testing is Inappropriate

Not all products can be tested to destruction. In such cases a simulation which traces the outcome of one or more hypothetical, risky events should be applied. A common approach is *scenario analysis,* in which one starts from a given event, then studies the different consequences which might evolve from it. Another approach, known as *failure modes and effects analysis,* systematically examines the failure modes of each component, without, however, focusing on causes or interrelationships among the elements of a complex system. In contrast to this is the *fault-tree analysis* method, where one proposes a system failure and then traces the event back to possible causes at the component level.

Of these several techniques, the fault-tree method can perhaps most effectively illustrate the disciplined approach required to capture as much as possible of everything that affects the proper functioning and safety of a complex system. Please note, however, that no safety analysis should be attempted without a thorough understanding of the physical aspects of the

FIGURE 4-6
A simple water system.

system under study; the mere use of ready-made computer programs does not suffice except to facilitate intermediate calculations.

To illustrate the use of the fault-tree technique we resort to a rather simple example, a water system without filtration plant, depicted in Fig. 4–6. A fault tree for that water system is shown in Fig. 4–7. We start with the system failure at the top and work down to failures in various subsystems, components, and outside factors or events which could have given rise to the problem. Each level in the tree lists events that could have caused the problem listed in the level above it. "Could have" implies that one *or* more events could be the cause for the event at the next higher level. Sometimes one *and* another event—perhaps several events—must all occur for that next event to happen. Thus there are two types of logical statements that appear on the chart, OR and AND.

The fault tree in Fig. 4–7 has not been completed. There are several further levels which could be indicated. There are also possible omissions at the levels shown. But even though incomplete, this fault tree can give us a good qualitative sense of the types of risks to which the water system is exposed. Some analysts proceed further and attach probability figures to each event. The accuracy of such figures is always problematic, however, particularly when there is the chance of common-mode failures. An earthquake, for example, could damage not only the reservoir—it could cause any of the events in Fig. 4–7 trailed by the circled E, eventually affecting delivery of water: Say an earthquake causes crumbling of river-banks and other damage leading to silting of the pump's source of water. If not properly filtered out, the silt could cause the impeller of the pump to wear out in no time at all. Alternatively the silting could cause the inlet to clog, thus stopping the inflow of water necessary to the pump's operation.

In the disaster that befell the passengers and the crew of the *Titanic*, the fact that the accident occurred at night certainly played its part as a "common-mode event" in producing the tragedy (Machol, 53). The iceberg could not be seen easily at night, the only radio operator on a nearby ship was asleep, and abandoning ship was more difficult at night than it would have been during the day. Thus because of the difficulty in foreseeing all common-mode events, one should treat the results of quantitative risk assessments with caution (see Study Question 4 below for a numerical example of this problem).

The strength of a fault-tree analysis lies in its qualitative aspects. It assists in the exposure of hitherto unforeseen situations. In a real-life water system analysis we would test for the availability of potable water and its usefulness. Water which is contaminated is not useful, even when available in large quantities at the householder's tap. Even high-quality water may not be useful when the sewer system is inoperative. On the other hand, even when no water comes out of the tap, water may still be available: from reserve tanks or from emergency delivery trucks. Thus we note that the fault-tree examination can be extended to embrace emergency measures as well.

FIGURE 4-7
Fault tree for the water system shown in Fig. 4-6.

Study Questions

1 "It is a sobering but uncertain possibility that our ability to respond to *unknown* hazards is diminished by the prevailing emphasis on control of the *known* and the specific" (Kates, *Preface,* emphasis added). Describe situations where this statement applies (for example, the past emphasis on fire hazards and the neglect of shock hazards in electrical wiring).

2 Testing is a critical step in product development and manufacture. It is not free from external influences, as the case described below reveals. Discuss how you would proceed with the testing program.

XYZ Aircraft is developing a new airplane. FAA procedures require a simulated emergency evacuation test in which the maximum number of people expected to occupy the aircraft at one time must be evacuated from it in less than 5 minutes. The test is conducted on a prototype plane in a dark hangar. It will be very expensive (insurance costs alone are considerable), and it has attracted a great deal of attention from the FAA, airline clients, and the media. The problem is that the emergency door and inflatable slide system have already revealed some serious design problems. Attempts to solve those problems have also already caused the developmental budget to be exceeded. Only two months remain until the scheduled date of the test. If the plane fails it, XYZ will be greatly embarrassed, the FAA will be alerted to the problems, and an even larger financial burden will be placed on XYZ, what with the need to rerun the tests on top of the necessity still to solve the design problems themselves.

The possible actions that XYZ can take include the following:

a Expect the test to be successful. Construction of the test aircraft can proceed while design changes are made on the emergency systems. The financial savings will be great if the plane passes the test.

b Postpone the test. Allow more time for redesign. Avoid a possible rerun of the test at a high cost.

3 Draw a fault-tree diagram for the event "automobile passenger falls out of a car during accident."

4 This is a problem which involves some manipulation of logic operations and probabilistic data. It is introduced here to demonstrate that common cause events can drastically reduce a system's (a brake's) calculated reliability.

Examine first the fault tree shown in Fig. 4–8a. Let P_J be the probability that event J occurs; let $\bar{P}_J$ be the probability that event J does not occur, with $\bar{P}_J = 1 - P_J$. For simplicity, let $P_D = P_E = P_F = P_G = 10^{-6}$. Then $P_B = 1 - \bar{P}_D\bar{P}_E = P_D + P_E - P_D P_E \cong 2 \cdot 10^{-6}$. Similarly $P_C \cong 2 \cdot 10^{-6}$. *The top event A* will then occur with probability $P_A = P_B P_C = 4 \cdot 10^{-12}$.

Now assume that the failure of the plunger in the rear half of the cylinder is not independent of the failure of the plunger in the front half. For instance, both failures could have originated from a mismatch in the properties of the plunger rubber and the brake fluid, thereby weakening the rubber. If such is the case, a different fault tree needs to be drawn, as shown in Fig. 4–8b. What is the value of P_A now?

5 Prepare a list of about ten major types of building failures. Describe the failures

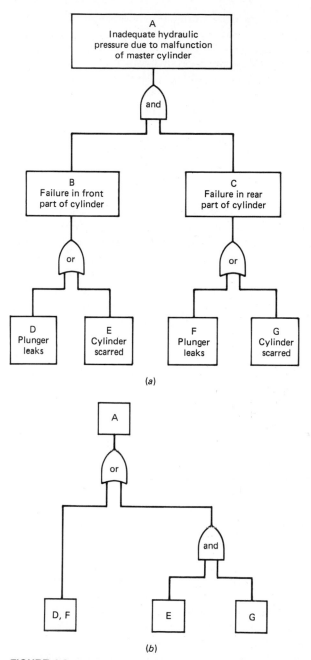

FIGURE 4-8
(a) Fault tree for a master brake cylinder of an automobile; (b) fault tree redrawn to account for a common-mode failure. *(Adapted from Hammer, pp. 220–221.)*

and give probable causes, such as faults in design, construction, or materials. Discuss which of those faults seems to be the most common. For assistance you may refer to accounts by the following authors listed in the Bibliography: Hayward; Klein and Lancaster; McQuade; Ransom (Chaps. 1 and 12). For a case study giving greater detail, consult Kardos. If you wish to look at dam failures, consult Jansen.

RISK-BENEFIT ANALYSES AND INDUCEMENTS TO REDUCE RISK

Many large projects, especially public works, are justified on the basis of a risk-benefit analysis. The questions answered by such a study are the following: Is the product worth the risks connected with its use? What are the benefits? Do they outweigh the risks? We are willing to take on certain levels of risk as long as the project (the product, the system, or the activity which is risky) promises sufficient benefit or gain. If risk and benefit can both be readily expressed in a common set of units (say, dollars), it is relatively easy to carry out a risk-benefit analysis and to determine whether or not we can expect to come out on the side of benefit. For example, an inoculation program may produce some deaths, but it is worth the risk if many more lives are saved by suppressing an imminent epidemic.

A closer examination of risk-benefit analyses will reveal some conceptual difficulties. Both risks and benefits lie in the future. Since there is some uncertainty associated with them, we should deal with their expected values; in other words, we should multiply the magnitude of the potential loss by the probability of its occurrence, and similarly with the gain. But who establishes these values, and how? If the benefits are about to be realized in the near future but the risks are far off (or vice versa), how is the future to be discounted in terms of, say, an interest rate so we can compare present values? What if the benefits accrue to one party and the risks are incurred by another party?

If either or both benefits and risks are comprised of ingredients which cannot be added in a common set of units, for example health, aesthetics, and dollars, how should one proceed? At most one could compare designs which satisfy some constraints in the form of "dollars not to exceed X, health not to drop below Y" and attempt to compare aesthetic values with those constraints. Or when the risks can be expressed and measured in one set of units (say, deaths on the highway) and benefits in another (speed of travel), we can employ the ratio of risks to benefits for different designs in comparing them with each other.

It should be noted that risk-benefit analysis, like cost-benefit analysis, is concerned with the advisability of undertaking a project. When we judge the relative merits of different designs, however, we move away from this concern. Instead we are dealing with something similar to cost-effectiveness analysis, which asks what design has the greater merit given that the project is actually to be carried out. Sometimes the shift from one type of consideration to the other is so subtle that it passes unnoticed. Nevertheless, engineers should be aware of the differences so that they do not unknowingly carry the assumptions behind one kind of concern into their deliberations over the other.

These difficulties notwithstanding, there is a need in today's technological society for some commonly agreed upon process—or at least a process open to scrutiny and open to modification as needed—for judging the acceptability of potentially risky projects. What we must keep in mind is the following ethical question: *"Under what conditions, if any, is someone in society entitled to impose a risk on someone else on behalf of a supposed benefit to yet others?"* (Council for Science and Society, 37). In examining this problem further, we can trace our steps back to an observation on risk perception made earlier: A risk to a known person (or to identifiable individuals) is perceived differently by people than statistical risks merely read or heard about. Engineers do not affect just an amorphous public— their decisions have a direct impact on people who feel the impact acutely, and that fact should be taken into account equally as seriously as studies of statistical risk.

Personal Risk

Given sufficient information, an individual is able to decide whether or not to participate in (or consent to be exposed to) a risky activity (an experiment). Chauncey Starr has prepared some widely used figures which indicate that individuals are more ready to assume voluntary risks than they are to be subjected to involuntary risks (or activities over which they have no control), even when the voluntary risks are 1000 times as likely to produce a fatality as are the involuntary ones. We show this graphically in Fig. 4–9.

The difficulty in assessing personal risks arises when we consider those that are involuntary. Take John and Ann Smith and their discomfort over living near a refinery. Assume the general public was all in favor of building a new refinery at that location, and assume the Smiths already lived in the area. Would they and others in their situation have been justified in trying to veto its construction? Would they have been entitled to compensation if the plant was built over their objections anyway? If so, how much compensation would have been adequate? These questions arise in numerous instances. Nuclear power plant siting is another example. Indeed, Fig. 4–9 was produced in the context of nuclear safety studies.

The problem of quantification alone raises innumerable problems in

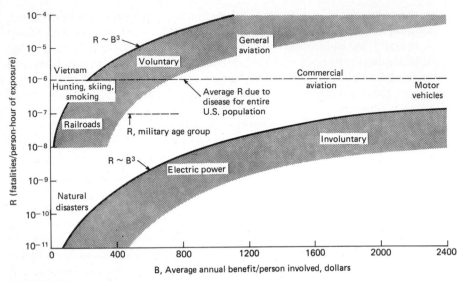

FIGURE 4-9
Willingness to assume voluntary risks as opposed to involuntary ones correlated to benefits those risks produce. *(After Starr, 1969, p. 1234.)*

assessing personal safety and risk, as was alluded to earlier. How, for instance, is one to assess the dollar value of an individual's life? This question is as difficult as deciding whose life is worth saving, should such choice ever have to be made.

Some would advocate that the marketplace should decide, assuming market values can come into play. But today there is no over-the-counter trade in lives. Nor are even more mundane gains and losses easily priced. If the market is being manipulated, or if there is a wide difference between "product" cost and sales price, it matters under what conditions the buying or selling takes place. For example, if one buys a loaf of bread, it can matter whether it is just one additional daily loaf among others one buys regularly or whether it is the first loaf available in weeks. Or if you are compensated for a risk by an amount based on the exposure tolerance of the "average" person, yet your tolerance of a condition or your propensity to be harmed is much greater than average, the compensation is apt to be inadequate.

The result of these difficulties in assessing personal risk is that analysts employ whatever quantitative measures are ready at hand. In regard to voluntary activities one could possibly make judgments on the basis of the amount of life insurance taken out by an individual. In assessing a hazardous job one might look at the increased wages a worker demands to carry out the task. Faced with the wide range of variables possible in such assessments, one can only suggest that an open procedure, overseen by trained arbiters, be employed in each case as it arises. On the other hand, for people taken in a population-at-large context, it is much easier to use statistical averages without giving offense to anyone in particular. The ethical implications of that practice will be addressed in the following subsections.

Public Risk and Public Acceptance

Risks and benefits to the public at large are more easily determined because individual differences tend to even out as larger numbers of people are considered. The contrast between costs of a disability viewed from a private value system and from a societal value system, for example, is vividly illustrated in Fig. 4–10. Also, assessment studies relating to technological safety can be conducted more readily in the detached manner of a macroscopic view as statistical parameters take on greater significance. In that context, the National Highway Traffic Safety Administration (NHTSA) has proposed a value for human life based on loss of future income and other costs associated with an accident. Intended for study purposes only, NHTSA's "blue book value" amounted to $200,725 in 1972 dollars. This is certainly a more convenient measure than sorting out the latest figures from court cases. (On April 23, 1981, for example, the *Los Angeles Times* reported settlements for relatives of victims of a 1979 DC-10 crash in Chicago at $2,287,257 for a 36-year-old promising executive, $750,000 for

FIGURE 4-10
Value systems for social costs of disability. Using the National Safety Council equivalent of 6000 disability days for death and L.A. Sagan's 1972 assumed rate of $50 per day of disability (Sagan, 488) yields a death of equivalent of $300,000—valid for societal value analysis only. *(After Starr, Rundman, and Whipple, p. 637.)*

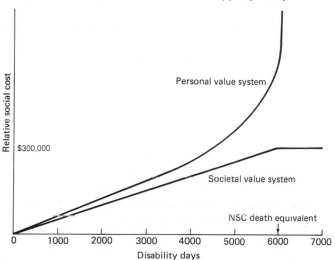

a telephone company employee, and $275,000 for a stewardess on duty.)

NHTSA, incidentally, emphasized that "placing a value on a human life can be nothing more than a play with figures. We have provided an estimate of some of the quantifiable losses in social welfare resulting from a fatality and can only hope that this estimate is not construed as some type of basis for determining the 'optimal' (or even worse, the 'maximum') amount of expenditure to be allocated to saving lives" (O'Neill and Kelley, 30).

Accounting Publicly for Benefits and Risks

The conclusions of risk assessment and cost-benefit studies are increasingly being challenged by special interest groups. And though engineers are generally reluctant to face the rough and tumble of the political and legal arenas, they are often called upon as expert witnesses in such cases. When testifying, they will find that they are treated with far more respect than perhaps equally well-versed lay people who, as interested citizens, have volunteered their time to study a crucial issue. But if formal expertise bestows a certain power on engineers, it behooves them not to abuse that power. There is a *noblesse oblige* which demands that they remain as objective as humanly possible in their investigations and the conclusions they reach and that, in order to place their testimony in the proper perspective, they state at the outset of testifying any personal biases they may have about the subject at issue.

No expert—or even group of experts—can be expected to be omniscient. Hence the public processes designed to establish safeguards and reasonable regulations in relation to technology themselves suffer from the already mentioned problem of incomplete knowledge that engineering is subject to. Moreover, in the view of a judge who has heard many cases involving new technologies and who has written searchingly on the subject, there is yet another problem affecting public accountability for risk:

> The other kind of uncertainty that infects risk regulation comes from a *refusal to face the hard questions created by lack of knowledge*. It is uncertainty produced by scientists and regulators who assure the public that there are no risks, but know that the answers are not at hand. Perhaps more important, it is a false sense of security because the hard questions have never been asked in the first place.
>
> In the early days of nuclear plant licensing, for example, the problem of long-term waste disposal was never even an issue. Only after extensive prodding by environmental and citizens' groups did the industry and regulators show any awareness of waste disposal as a problem at all. Judges like myself became troubled when those charged with ensuring nuclear safety refused even to recognize the seriousness of the waste disposal issue, much less to propose a solution (Bazelon, 279, emphasis added).

A willingness to admit uncertainty and bias and to reveal methodology and sources is particularly important when numerical data and statistics are presented. We have already mentioned how conceptions of risk can vary—even be turned around—depending on how the facts are presented. How presentations of data interpretation (even the best intended) can be misleading is pointed out by W. Hammer in his discussion of tabular material we give here in Table 4–1. The table presents statistics on accident fatalities for the U.S. population. The approximate individual risk entries were calculated as part of a study for the U.S. Atomic Energy Commission in 1974 (Rasmussen). Hammer writes:

The values [in Table 4–1] illustrate one problem with the use of quantitative assessments. All the fatality risk values shown are based on total U.S. population in 1969, which may not be valid in some cases. For example, if half the people of the United States do not travel by air, the probability of any one of them being killed (or of having been killed in 1969) in a plane crash is zero. The probability

TABLE 4-1
INDIVIDUAL RISK OF ACUTE FATALITY BY VARIOUS CAUSES

Accident type	Total number for 1969	Approximate individual risk of acute fatality, probability/yr.*
Motor vehicle	55,791	3×10^{-4}
Falls	17,827	9×10^{-5}
Fires and hot substance	7,451	4×10^{-5}
Drowning	6,181	3×10^{-5}
Poison	4,516	2×10^{-5}
Firearms	2,309	1×10^{-5}
Machinery (1968)	2,054	1×10^{-5}
Water transport	1,743	9×10^{-6}
Air travel	1,778	9×10^{-6}
Falling objects	1,271	6×10^{-6}
Electrocution	1,148	6×10^{-6}
Railway	884	4×10^{-6}
Lightning	160	5×10^{-7}
Tornadoes	91	4×10^{-7}
Hurricanes	93	4×10^{-7}
All others	8,695	4×10^{-5}
All accidents		6×10^{-4}

*Based on total U.S. population.
Source: Rasmussen, p. 230.

of the average air traveler being killed is increased to 1.8 times 10^{-5}. The probability of a person who travels more than the average (which isn't specified) is even higher. This method of risk assessment becomes even more invalid when the operation considered is one in which few persons participate. Assume that there are 160 persons killed in hang-gliding accidents in a year (the same as . . . [the number] killed by lightning). When the total U.S. population is used as the base for determining the risk, the probability of a fatality is 5 times 10^{-7}, as shown; a relatively safe operation. However, if there are only 20,000 enthusiasts who participate in hang gliding and they suffer a 160-person loss each year, the fatality risk is 8 times 10^{-3}. To make correct and comparable risk assessments it is therefore necessary to base them on correct and acceptable assumptions and data (Hammer, 246).

In another example, the automobile manufacturer Volvo advised the U.S. General Accounting Office not to rely on industry-generated cost information for safety features because "the auto industry has in some instances taken advantage of the lack of methodology and released biased material aimed purely at resisting regulation" (quoted in O'Neill and Kelley, 33).

Engineers should be concerned about numbers in assessing safety and risk, but they should insist on meaningful numbers. Accordingly they should regard statistics with caution, whatever source may have issued them. Engineers should also recognize the previously mentioned difficulties of measuring risks and benefits on a cardinal scale (that is, in absolute terms), and should instead employ ordinal rankings. One of the difficulties with risk-benefit and cost-benefit analyses, again, is the matter of who does the assessing. The parties that will be affected by a project are rarely polled, especially when they are not represented by an influential lobby or trade organization (Nelson and Peterson, 2).

But difficulties in publicly accounting for risks and benefits are not related only to methods of quantification. There is also the question of justice, which involves *qualitative* value judgments:

> There are things which are wrong to do regardless of the benefits of the consequences. This is contrary to the basic assumptions in cost-benefit analysis. Without going so far as to say that consequences are never important, we can say that they are not as important as the [analyses] would imply. . . . The type of action one does should be morally evaluated regardless of its consequences; if it is wrong to violate certain rights, then figuring out the benefits of the consequences of doing so is irrelevant (Nelson and Peterson, 4).

Paying compensation, for example, may be an efficient and bureaucratically pleasing process for trying to make restitution when harm has been done, and indeed it is an improvement over earlier government and business procedures. But efficiency in itself does not promote ethics, as much as one may be tempted to equate the two. As an example we may cite the attempts of San Francisco in the 1960s to demolish in one fell swoop a large area slated for urban renewal. Even if the residents being displaced could have afforded to live in the new housing units once built, they would have scattered and the neighborhood would have been destroyed by the time the new dwellings were completed. The response by those residents was a near revolt which caused the city to rethink its cost-benefit formula and to set about rebuilding the area on a block by block basis, with accommodations for the temporarily displaced being provided by the city. A more expensive solution, certainly, but also more humane. And engineers need to be sensitive to such considerations.

Incentives to Reduce Risk

The engineer is faced with the formidable tasks of designing and manufacturing safe products, of giving a fair accounting of benefits and risks in regard to those products, and of meeting production schedules and helping his or her company maintain profits all at the same time. Of these objectives, product safety should command top priority. Yet this is not often so in practice, partly because of some commonly held misconceptions which mitigate against applying the extra thought and effort required to make a product safe.

Among the popular though faulty assumptions about safety are the following (adapted from Hammer, 52).

Assumption: Operator error and negligence are the principal causes of all accidents.

Reality: Accidents are caused by dangerous conditions that can be corrected.

For example, introduction of automatic couplers drastically reduced the number of deaths and injuries suffered by train workers. Dangerous design characteristics of products cause more accidents than failures (by fatigue, etc.) of components.

Assumption: Making a product safe invariably increases costs.

Reality: Initial costs need not be higher if safety is built into a product from the beginning. It is design changes at a later date that are costly. Even then life cycle costs can be lower for the redesigned, safe product.

Assumption: We learn about safety after a product has been completed and tested.

Reality: If safety is not built into the original design, people can be hurt during the testing stage.

Reluctance to change a design may mean safety features will not be incorporated into the product.

Assumption: Warnings about hazards are adequate; insurance coverage is cheaper than planning for safety.

Reality: A warning merely indicates that a hazard is known to exist, and thus provides only minimal protection against harm. Insurance rates are skyrocketing. Recall actions can add to costs, even when no accidents have occurred.

Engineers should recognize that reducing risk is not an impossible task, even under financial and time constraints. All it takes in many cases is a different perspective on the design problem: a recognition from the outset that one is embarking on an experiment in which safety is an important factor.

Some Examples of Improved Safety

This is not a book on design; therefore only a few simple examples will be given to show that safety need not rest on elaborate contingency features.

The first example is the magnetic door catch introduced on refrigerators to prevent death by asphyxiation of children accidentally trapped in them. The new catch permits the door to be opened from the inside without major effort. It also happens to be cheaper than the older types of latches.

The second example is the dead-man-handle used by the engineer (engine driver) to control a train's speed. The train is powered only as long as some pressure is exerted on the handle. If the engineer becomes incapacitated and lets go of the handle, the train stops automatically. Perhaps cruise-controls for newer model automobiles should come equipped with a similar feature.

Railroads provide the third example as well. Old-fashioned semaphors actuated by cable usually indicated STOP when the arm was lowered. This was the position the arm assumed all by itself if the cable snapped accidentally. Here we have an early instance of a fail-safe design dating back more than a hundred years.

The motor-reversing system shown diagrammatically in Fig. 4–11 is still another example where the introduction of a safety feature involves merely the proper arrangement of functions at no additional expense. As the mechanism is designed in Fig. 4–11a, sticky contacts could cause battery B to be shorted, thus making it unavailable for further use even after the contacts are coaxed loose. A simple reconnection of wires as shown in Fig. 4–11b removes that problem altogether.

As a final example we mention the Volkswagen safety belt. A simple attachment on the door ensures that the belt automatically goes into place whenever one enters the car. Forgetting to strap oneself in will no longer be a problem.

In the rush to bring a product onto the market, safety considerations are frequently slighted. This would not be so much the case if the venture were

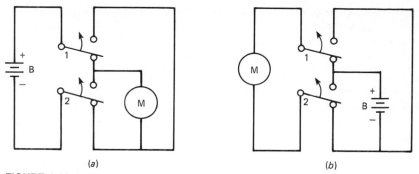

(a)   (b)

FIGURE 4-11
Motor-reversing system. (a) Arms 1 and 2 of the switch are both raised by a solenoid (not shown). If either one does not move—say, a contact sticks—while the other does, there is a short across the battery. The battery will discharge and be useless even after the trouble is detected. (b) By exchanging the positions of battery and motor, a stuck switch will cause no harm to the battery. (The motor can be shorted without harm.)

regarded as an experiment—an experiment which is about to enter its active phase as the product comes into the hands of the user. Space flights are carried out with such an attitude, but commercial ventures involve less obvious dangers and therefore less attention is usually paid to safety. If moral concerns alone do not sway engineers and their employers to be more heedful of potential risks, then recent trends in product liability should certainly do so.

Liability

The last two decades have seen a drastic change in legal protection for the consumer. Richard C. Vaughn, in his informative book *Legal Aspects of Engineering*, describes the changes as follows:

> Early English social and legal philosophy reflected the manufacturing nature of the economy. Producers of goods and services were held in high esteem. Their success meant success of the nation. The legal climate fostered their growth. Both logic and social philosophy supported the legal defense available if someone complained about a product—*caveat emptor* (the buyer beware). The logic was simple: one should examine what he is to receive before he buys it. If he is so negligent that he does not examine before he buys, then he should live with his bad bargain. Legal support of an action to recover for a bad product would be, in effect, support of buyer negligence and the law usually will not aid those who are negligent. But then, of course, the products produced in those days were somewhat more easily examined than what we buy today. . . .
>
> Another defense in the producers' armament was "privity of contract"—the idea that one who is not a party of a contract should have no rights arising from it. In other words, if one was injured by a product but he did not buy it directly from the manufacturer, he could not act against that manufacturer to recover for his

injury. The producer or manufacturer only needed to interpose a middleman—a wholesaler or retailer—as an insulator. Then, if the injured person could prove he was the buyer of the product, he might sue the middleman, but he could not reach "the deep pocket" (Vaughn, 41).

A turning point in the reversal of these attitudes came in 1916 when Judge Cordoza found the Buick Motor Company responsible for the injuries suffered by one McPherson when a wheel on a new Buick collapsed. Gradually thereafter, over a period of half a century, the notion took hold that a manufacturer can be held liable for injuries resulting from negligence in a product's design or manufacture. Still, such negligence had to be proven. Then, in 1963, the concept of strict liability was established in California by the case Greenman vs. Yuba Power Products, a concept soon thereafter incorporated into tort law by most states. Strict liability means it is sufficient for a product to have been defective as sold for the manufacturer to be held liable for any harm that results to users. Negligence is not at issue. What matters is that the product has a defect not obvious to users.

Despite the responsibilities implied by the doctrine of strict liability, negligence certainly remains a more grievous offense, as does breach of warranty. The latter is interpreted as a breach of contract and usually covers only the purchaser and members of the purchaser's household, not just any user in general. The warranty is established by advertising, labels, and other information which causes the buyer to expect a serviceable and safe product.

Engineers—and students of engineering—need to be aware of strict liability. As Richard Moll writes: "The fact that proof of negligence is not essential to impose liability is a frightening prospect for most manufacturers. . . . The significance of the strict liability doctrine, as far as engineers are concerned, is that although in many cases it is impossible to test every product, the engineer must weigh the chances of a defect causing serious injury against the cost of eliminating or minimizing defects in the product" (Moll, 331).

While most engineers strive to be responsible in their work and have the safety of the public in mind, conditions imposed by employers may circumscribe these professional goals. On the other hand, they can also be sued individually, even when acting according to guidelines set by their employers. This may happen when an injured party is frustrated by laws which shield the employer or limit the employer's liability, as is the case with government agencies. Thus it happened that one county highway engineer was sued for damages exceeding $2 million. As Dean M. Dilley reports, "The plaintiff suffered permanent injuries in an automobile accident caused by a washed out section of highway, and the suit argued that the engineer's failure to repair the road gave rise to an action for negligence against him personally" (Dilley, 12).

Some companies and government agencies protect their engineers by allowing themselves to be sued for money damages over harm arising from activities of employees working within the scope of their professional duties. Engineers not so protected can resort to malpractice insurance. Independent engineers who contract for their work can minimize the effects of adverse legal judgments by writing liability limits into their contracts. This, in turn, can reduce their malpractice insurance rates. As has been mentioned before, engineering practice should be preventive or defensive in approach. A knowledge of liability is therefore well advised. Good introductory material is offered in books by Richard C. Vaughn (already mentioned) and by James F. Thorpe and William H. Middendorf (see the Bibliography).

On a larger scale, liability limitation was offered to electric utilities through the Price-Anderson Act in order to entice them into nuclear power ventures during the Atoms for Peace Program of the Eisenhower years. The $400 million limit seemed inadequate even then, but it was probably assumed that the government would step in with disaster aid in case of a serious accident. Unfortunately the expectation of bailouts with such aid has often resulted in inadequate planning by public agencies for potential disasters of many different types, from floods to droughts, from earthquakes to conflagrations. While the Three Mile Island Nuclear Power Plant incident of 1979 did not constitute a disaster in the usual sense, it serves as an example of the need to take safety seriously in large-scale engineering design, so it will be discussed in greater detail in the next section.

Study Questions

1 A 250,000-ton tanker (capable of holding 200,000 tons of oil) is cruising at 16 knots when engine power (25,000 hp) is lost. Control over the rudder is lost as well.

   a Estimate how long it will take in time and distance before the ship comes to a stop. Assume it had been on a straight course before the mishap. (Under normal circumstances, with reversing power available, it will take 3 miles or 22 minutes.)

   b What might be the consequences of the tanker running aground or colliding with another vessel?

   c How many tankers were lost at sea during the last decade? You may consult an almanac.

   d What are the benefits of operating a supertanker? Some are larger than 500,000 tons. Some 1-million-ton tankers have been planned. (An interesting source on this topic is Noel Mostert's book, *Supership*.)

2 The following appeared in the July 10, 1981, issue of *Science*:

The Supreme Court, in the cotton dust decision on 17 June, says explicitly that OSHA must ignore the results of any cost-benefit comparison when setting a standard for worker exposure to a hazardous substance. Justice William Brennan, writing for the court's five-person majority, said that "Congress *itself* decided the basic relationship between costs and benefits by placing the 'benefit' of the worker's health above all other considerations" when it wrote the law in 1970. Yet the agency cannot require exposure controls that are impossible to achieve, nor can it bankrupt an entire industry, Brennan wrote. He concluded that consideration of anything besides these questions would be inconsistent with Congress's direction (R. J. Smith, 185).

Discuss this opinion of the court and how you think it might affect efforts to combat byssinosis (brown lung disease suffered by textile workers). What interpretations do you think OSHA, the textile industry, and the textile workers' unions would give this court decision? If you were an engineer in a textile plant, how would you react? Discuss your reactions in terms of their ethical foundations.

3 The following excerpt is from an article by T. W. Lockhart, "Safety Engineering and the Value of Life." Attempt to answer the question posed by Lockhart. Try again after you have finished reading this entire chapter.

. . . there is an honored tradition in moral philosophy, associated primarily with Immanuel Kant, according to which human beings have a worth that is not commensurable with that of mere objects. According to this view, because of this incommensurability we must recognize and respect the liberty and dignity of each person and refrain from treating him merely as a means to some end. Human beings may not be used in order to achieve some higher good, for there is no higher good. Let us call this view the Incommensurability Principle.

The Incommensurability Principle has had a powerful appeal for many. This has been true mainly because it has been felt that unless it, or something like it, is accepted it is not possible to account for such fundamental human rights as the right not to be killed, the right not to have one's liberty abridged without just cause, and the right to be treated fairly and honestly. The Incommensurability Principle is clearly incompatible with an attempt to place a monetary value on human life or to justify actions on the basis of such a valuation. There is thus further reason for doubting the wisdom of any such attempt. . . .

Is it possible to reconcile the Incommmensurability Principle with the commonsense view that considerations of safety must be weighed against economic costs? (Lockhart, 3)

4 During the early stages of its development, crashworthiness tests revealed that the Pinto, a Ford automobile, could not sustain a front-end collision without the windshield breaking. A quick-fix solution was adopted: The drive train was moved backward. As a result the differential was moved very close to the gas tank. Thus many gas tanks collapsed and exploded upon rear-end collisions once the Pinto became available to the public. An account of the windshield episode is available in the literature (Camps, 119–129). Pinto explosions and resultant lawsuits have been widely reported in the press. Study the Pinto story and write it up as a case study. You may use the Predicasts F&S (Funk and Scott) Index of Corporations and Industries (Predicasts, Inc., Cleveland, Ohio) to find suitable references.

5 The U.S. Consumer Product Safety Commission states the following in its 1981 annual report:

Each year, an estimated 36 million Americans are injured and 30,000 are killed in accidents involving consumer products. The cost of these accidents is staggering—over $12.5 billion annually in medical costs and lost earnings.

These figures have been repeatedly cited in the literature to impress on engineers the need to design safe products. Review the data presented by the Commission in its annual and monthly reports and estimate what portion of the figures can be rightfully ascribed to poor design and what portion to consumer carelessness.

6 Only about 10 years ago 600 women reportedly lost hands (one, not both, we hope) in top loading washing machines every year (Papanek, 75). If you had been in the business of manufacturing washing machines then, how do you think you might have reacted and should have reacted to such statistics?

7 A worker accepts a dangerous job after being offered an annual bonus of $2000. The probability that the worker may be killed in any one year is 1/10,000. This is known to the worker. The bonus may therefore be interpreted as a self-assessment of life with a value equal to $2000 divided by 1/10,000, or $20 million. Is the worker more or less likely to accept the job if presented with the statistically identical figures of a $100,000 bonus over 50 years (neglecting interest) and a 1/200 probability of a fatal accident during that period?

8 As emerging technologies have reduced risks from some sources, other risks have frequently become prominent. For instance, efforts to combat famine by drawing and transporting water for irrigation brought early benefits coupled with lasting problems to such diverse areas as Mesopotamia and the Andes in the past, or to the Punjab and the Sahel in the present. We also observe that the decrease in infectious disease was accompanied by an increase in chronic disease. Give another similar example and discuss it as a problem in risk-benefit analysis.

9 The owner of a television set brought legal action against its manufacturer seeking compensatory and punitive damages for severe burns and other injuries suffered when the set burst into flames. It was revealed at the trial that other sets made by the manufacturer had also gone up in smoke and flames. Study the case as reported in the *Federal Reporter* and prepare a case study for use in discussing the legal and ethical implications of the manufacturer's actions (or lack thereof). The case is Zora H. Gillham vs. The Admiral Corporation. It was brought to the authors' attention by Donald E. Wilson. It appears in the *Federal Reporter*, 2d series, Vol. 253, F.2d, West Publishing Co., St. Paul, Minnesota, 1976.

THREE MILE ISLAND

Walter Creitz, president of Metropolitan Edison, the power company in the Susquehanna Basin, was obviously annoyed by a series of articles in the *Record*. This local, daily newspaper of York, Pennsylvania, had cited unsafe conditions at Metropolitan Edison's Three Mile Island Nuclear Power Plant Unit 2. Creitz dismissed the stories as "something less than a patriotic act—comparable in recklessness . . . to shouting 'Fire!' in a crowded theater." A few days later a minor malfunction in the plant set off a series of events which made "Three Mile Island" into household words across the world (Rogovin, 3).

A Chronology of Events

Very briefly the events were as follows (Ford; Mason; Rogovin). At 4 A.M. on March 28, 1979, Unit 2 (Fig. 4–12) was operating normally under full automatic control at 97 percent of its rated power output. For about 11 hours a maintenance crew had been dealing with a routine minor problem. Normally resin, which functions to purify water in the secondary coolant system, is automatically flushed out of the demineralizer where it is used—No. 14 in Fig. 4–12. Yet enough resin had accumulated to block the pipe leading from the demineralizer to a tank where it is regenerated before being recycled. While working, the maintenance crew apparently caused water to back up in the pipe, which then caused isolation valve 13 to drift toward the closed position.

With water flow interrupted in the secondary loop 26, all but one of the condensate booster pumps turned off. That caused the main feedwater pumps 23 and the turbine 10 to shut down as well. In turn, an automatic emergency system started up the auxiliary feedwater pumps 25. But with the turbines inoperative, there was little outlet for the heat generated by the fission process in the reactor core. The pressure in the reactor rose to over 2200 pounds per square inch, opening a pressure relief valve 7 and signaling a SCRAM in which control rods are lowered into the reactor core to stop the main fission process from continuing.

The open valve succeeded in lowering the pressure, and the valve was readied to be closed. Its solenoid was de-energized and the operators were so informed by their control-panel lights. But something went wrong: The valve remained open, contrary to what the control panel indicated. Apart from this failure everything else had proceeded automatically as it was supposed to. Everything, that is, except for one other serious omission: The pumps 25 could not supply the auxiliary feed water because block valves 24 had inadvertently been left closed after maintenance work done on them two days earlier. Without feedwater in loop 26 the steam generator 3 boiled dry. Now there was practically no heat removal from the reactor, except through the relief valve. Water was pouring out through it at the rate of 220 gallons per minute. The reactor had not yet cooled down, and even with the control rods shutting off the main fission reaction there would still be considerable heat produced by the continuing radioactive decay of waste products.

Loss of water in the reactor caused one of a group of pumps, indicated as No. 15 in Fig. 4–12, to start automatically; another pump was started by the operators to rapidly replenish the water supply for the reactor core. Soon thereafter the full emergency core cooling system went into operation in response to low reactor pressure. Low reactor pressure can promote the formation of steam bubbles which reduce the effectiveness of heat transfer from the nuclear fuel to the water. There is a pressurizer which is designed to keep the reactor water under pressure. (The relief valve sits atop this pressurizer.) The fluid level in the pressurizer is also used as an indirect, and only, means of measuring the water level in the reactor.

The steam in the reactor vessel caused the fluid level in the pressurizer to rise. The operators, thinking they had resolved the problem and that they now had too much water in the reactor, shut down the emergency core cooling system and all but one of the emergency pumps. Then they proceeded to drain water at a rate of 160 gallons per minute from the reactor, causing the pressure to drop. At this point they were still unaware of the water escaping through the open relief valve. Actually they assumed some leakage, which occurred because of poor valve-seating even under normal circumstances. It was this which made them disregard the high-temperature readings in the pipes beyond 7.

The steam bubbles in the reactor water covered much of the fuel, and the tops of the fuel rods began to crumble. The chemical reaction between the steam and zircaloy covering the fuel elements produced hydrogen, some of which was released into the containment structure.

The situation was beginning to get really serious when, 2 hours after the initial event, the next shift arrived for duty. With some fresh insights into the situation, the relief valve was deduced to be open. Blocking valve 9 in the relief line was then closed. Soon thereafter, with radiation levels in the containment building rising, a general alarm was sounded. While there had been telephone contact with the Nuclear Regulatory Commission (NRC), as well as with Babcock & Wilcox (B&W), who had constructed the reactor facility, no one answered at NRC's regional office and a message had to be left with an answering service. The fire chief of nearby Middletown was to hear about the emergency on the evening news.

In the meantime, a pump was transferring the drained water from the main containment building to the adjacent auxiliary building, but not into a holding tank as intended; because of a blown rupture disk, the water landed on the floor. Eventually there was to be sufficient airborne radiation in the control room to force evacuation of all but essential personnel. Those left behind wore respirators, making communication difficult.

Eventually the operators decided to turn the high-pressure injection pumps back on again, as the automatic system had been set to do all along. The core was covered once more with water, though there were still some steam and hydrogen bubbles on the loose. Thirteen and one-half hours after the start of the episode there was finally hope of getting the reactor under

FIGURE 4-12
Schematic diagram of Three Mile Island Nuclear Power Plant Unit 2. (*Adapted from IEEE Spectrum, November 1979, copyright* © 1979 by the Institute of Electrical and Electronics Engineers, Inc., and from Rogovin and Frampton.)

control. Confusion over the actual state of affairs, however, continued for several days.

Nationwide the public watched television coverage in disbelief as responsible agencies displayed their lack of emergency preparedness as they were groping for the next moves at the reactor site and at evacuation planning centers. Now, several years later, the decommissioning of the reactor has been temporarily halted while negotiations are underway to decide how the enormous costs of this fiasco—a billion dollars and above—are to be shared. A major physical catastrophe had been averted—the containment structure had served its purpose—but in terms of economic losses and shaken public confidence the accident could be called a disaster. Three Mile Island has become a household term and the topic of many books (see the Bibliography under Bill Keisling, Daniel Martin, Mark Stephens, and Mike Gray and Ira Rosen). Now it behooves us to draw the proper lessons from the incident.

Lessons from Three Mile Island

The March 1979 incident at Three Mile Island (TMI) underscores a number of lessons which experience with other large projects has taught the engineering profession. First is the reminder that Murphy's law is not to be taken lightly: If anything can go wrong, it will. Second is the observation that planning for emergencies must precede a crisis and cannot be left until a crisis is upon us. Third is the fact that engineers who raise serious complaints can have good reasons for doing so. Let us elaborate on this last point by mentioning the experiences of a few engineers who before the TMI events had expressed concerns about the type of B&W reactors used there.

1 Stephen Hanauer, a government nuclear safety expert, communicated in 1969 to then Atomic Energy Commission head Glenn Seaborg his concern regarding common-mode failure possibilities at nuclear power plants. Later he became particularly worried about the poor performance record of valves in nuclear power plants. He also decried the lack of proper analysis of field reports which would help identify weak spots in design (Ford, April 6, 1981, 49).

As it turned out, the largest single failure at TMI was of a common-mode type, although not of a physical nature, as is more usual. It lay in the operators' inadequate training. No one on duty at the plant that early morning of March 28 was a nuclear engineer, and no one was even a college graduate; no one in charge was trained to handle complex reactor emergencies. [Interestingly enough, the 1974 Rasmussen Report on nuclear power plant safety (see the Bibliography) had singled out human response as the weakest link in plant reliability.]

Moreover, Hanauer's concern over valves was well founded. Problems with valves have not been restricted to the nuclear industry, and the particular type of relief valve that malfunctioned at TMI had done so earlier at other nuclear power plants. In fact, since 1970 eleven pilot-operated relief valves had stuck open at other such plants (Lombardo, 55).

2 John O'Leary, Deputy Secretary of Energy when he presided at the opening ceremonies of TMI's Unit 2, had earlier prepared a memo for incoming President Carter. Then a deputy director for licensing at the Atomic Energy Commission, he had written that "the frequency of serious and potentially catastrophic nuclear incidents supports the conclusion that sooner or later a major disaster will occur at a major generating facility" (Ford, April 6, 1981, 49). Yet the probability of serious accidents continued to be given such a low value that it was not thought necessary to plan for orderly emergency measures involving off-site populations and agencies.

3 James Creswell was a reactor inspector for the NRC assigned to the startup of Toledo Edison's Davis-Besse nuclear generating station, which also used a reactor built by B&W. Reading a B&W report on some strange behavior of its unit at Rancho Seco in California following the accidental dropping of a small light bulb into the main control panel, Creswell noticed startling similarities to an incident which had occurred at Davis-Besse during low-power tests in September 1977. A severe and sudden increase in heat had taken place in both cases, at Rancho Seco because of faulty control signals produced by difficulties with the control panel and at Davis-Besse because of failure of the main feedwater system. At both plants the instruments had not given the operators adequate indication of the reactors' true operating conditions. At Davis-Besse the pressure relief had stuck open and the operator had misinterpreted the level of water in the reactor based on indications of the level in the pressurizer. This had led them to mistakenly shut off the emergency core cooling pumps, even at Davis-Besse. The only difference with TMI was that Davis-Besse had been operating at 9 percent of rated capacity (as against 96 percent at TMI) and the failure of the relief valve to close had been detected after 22 minutes (instead of over 2 hours).

Creswell worried about this and, from early 1978 on, tried to communicate his concerns to various parties at the NRC, the utility, and B&W. Eventually he took a day off and flew at his own expense to Bethesda, Maryland. There he met with two NRC commissioners who were willing to listen. The result was a memo to the NRC staff requesting answers to some of Creswell's questions. The memo was delivered the day after the TMI incident.

While the similarities between what happened at Davis-Besse and at Rancho Seco are of interest, the near duplication at TMI of the occurrences at Davis-Besse makes a strong case for the free exchange of information and expeditious correction of faults, unfettered by organizational expediencies or short-range interests. And an ironic twist to the story points out clearly what approaches to safety are usually the more valued: Creswell eventually received a special NRC award of $4000 for his efforts. Two other officials, who had earlier been unresponsive to Creswell's pleas, were awarded $10,000 and $20,000 each for their efforts during the ensuing dramatic emergency at TMI.

4 Also late in 1977 Carlyle Michelson, a senior nuclear engineer with the Tennessee Valley Authority (TVA), had been questioning safety aspects of a new B&W reactor for TVA's Bellefonte Nuclear Plant. Included in his analysis was the possible formation of a steam bubble in the reactor's cooling-water system, something which had also been mentioned by Creswell. Michelson's study, which additionally mentioned the dangers of misreading the reactor coolant level by looking at the pressurizer, received some support on its way to the NRC's Division of System Safety. If handled according to protocol, the memo should have been sent on to the Division of Operating Reactors, whence reactor users, particularly at Davis-Besse, would have been alerted to the problems. The assistant director of the Division of System Safety later said that the Davis-Besse analysis was the responsibility of the Office of Inspection and Enforcement. He thought they would have taken action if any was needed (Ford, April 13, 1981, 82). Ten days after the TMI incident, the Advisory Committee on Reactor Safeguards asked the NRC to carry out some of Michelson's recommendations.

5 At B&W, meanwhile, the 1977 occurrence at Davis-Besse caused concern. Before the TMI incident the company did not have any formal procedure to analyze on-going problems at B&W-equipped plants or to study reports filed by its customers with the NRC. B&W was busy building and shipping new reactors. Nevertheless, the Davis-Besse incident was sufficiently unusual so that B&W engineers were sent out to investigate. Some individuals maintained their interest and tried—in vain—to get word out to B&W customers: John Kelly, a plant-integration engineer; Bert Dunn, manager of emergency-cooling-system analysis; and Donald Hallman, a customer service employee. Internal B&W memos later revealed the company's awareness of reactor defects. It denied NRC charges that it had failed to notify the Commission, but it instituted measures to make sure that this would not occur again (Ford, April 13, 1981, 92).

The reactor manufacturer's preoccupation with pressing production schedules, agency hearings, and the like, left little time for what we would call "monitoring of ongoing experiments." The NRC could similarly be faulted for launching new "experiments" without keeping sufficient track of those already underway. Nor is this tendency restricted to the nuclear industry. Everywhere we find the new venture more alluring, the high drama of the emergency more thrilling than the slow, persistent, preventive approach. It is up to professionals to make defensive engineering the accepted style.

Study Questions

1 Collect some examples of literature promoting and criticizing use of nuclear power. Are the statements factually complete? Do you find yourself agreeing with whatever item you have read last? Discuss the responsibility of experts to reasonably educate the populace about the issues involved, pro and con.

2 It has been said that Three Mile Island showed us the risks of nuclear power and the Arab oil embargo the risk of having no energy. Removing hazardous products or services from the market has been criticized as closing out the options of those with rising aspirations who can now afford them and who may all along have borne more than their share of the risks without any of the benefits. Finally, pioneers have always exposed themselves to risk. Without risk there would be no progress. Discuss this problem of "the risk of no risk." (Compare: Wildavsky.)

3 A number of engineers engaged in nuclear power plant work have expressed their concern over inadequate attention to safety. Some resigned first, then went public with their testimony; some spoke out and were fired; others spoke up but nevertheless retained their positions. Examine the literature about the cases listed below to see if there were any issues of ethical import involved. If so, what were they?

   a Carl Houston, 1970, welding superintendent, Stone & Webster (Houston)
   b Peter Faulkner, 1974, systems application engineer, Nuclear Services Corporation (Faulkner)
   c Dale G. Bridenbaugh, 1976, manager of performance evaluation and improvement; Richard B. Hubbard, 1976, manager of quality assurance; Gregory C. Minor, 1976, manager of advanced control and instrumentation; all with Atomic Power Division, General Electric Company, San Jose, California (Kaplan)
   d Ronald M. Fluegge, 1976, safety analyst; Demetrios L. Basdekas, 1976, reactor engineer; both with Nuclear Regulatory Commission (Congressional Record)
   e Robert D. Pollard, 1976, project manager, Nuclear Regulatory Commission (Friedlander, 1976).

WHY SHOULD ENGINEERS CARE ABOUT SAFETY?

Now that we have discussed the ethics of safety in engineering from technical and economic points of view, let us tie it in more directly to philosophical considerations.

At one level it is obvious enough why engineers should care about safety, at least if the "should" means "morally should." Engineers should care about safety because they have important moral obligations to do so. But why do they have these obligations? What is the ground or justification of the obligations?

In one way or another, safety is involved in most of the thorny issues connected with engineering ethics. Certainly it is the most pressing consideration involved in matters of whistleblowing, confidentiality, and the exercise of professional autonomy. In fact, it is perhaps only a slight exaggeration to say that whereas medical ethics centers on the role of promoting health within the bounds of patient autonomy, and legal ethics centers on the advocacy of clients' rights within the bounds set by law, engineering ethics takes as its primary focus the promotion of useful technological products within the bounds of safety considerations that protect the people affected by those products. We can hardly expect that an understanding of the philosophical foundations of engineers' safety obligations will settle every practical dispute by determining the precise extent of those obligations. Yet such an understanding may provide an enriched grasp of how disputes arise, and it may facilitate a better-organized inquiry into them.

There are at least two philosophical issues we must examine in connection with the safety obligations of engineers. One has to do with basing the special obligations of engineers on a normative ethical theory of correct action. The second concerns individual engineers and how they in particular acquire special safety obligations.

Applying Ethical Theory

In searching for the foundations of engineers' obligations concerning safety, we assume that those obligations are not themselves moral ultimates which cannot be related to more basic moral principles. This assumption seems acceptable, for morality is not generated anew in and of itself within engineering or the other professions. We take this much for granted when we subject the standards endorsed by professionals in any field to critical scrutiny in the light of fundamental principles of justice and respect for others.

Ideally we would like to identify the most comprehensive, valid, ethical theory which displays engineers' safety obligations as special instances of more general and fundamental moral obligations. And so we turn to the familiar philosophical theories which—each in its own way—build upon, refine, and structure ordinary moral experience.

Under *act-utilitarianism*, then, the special safety obligations of engineers would ultimately reduce to one general obligation: to act so as to maximize the good consequences for everyone affected by engineering projects and products. *Rule-utilitarianism*, on the other hand, would have engineers act according to those rules which if widely followed would produce the best consequences for everyone affected. *Duty-based theories* would ground the obligations of engineers in one or more basic principles of duty. And a *rights-based ethical theory* would emphasize how engineers' safety obligations are based on the requirement that people respect the moral rights of those affected by their work (as will be shown later).

Of course in engineering as in philosophy itself there will be disagreement over which, if any, of these theories is ultimately more correct. Such disagreement sometimes generates skepticism about the usefulness of theories within applied ethics. It might seem that there is no point in attempting to apply principles until the theoretical disputes over the principles are resolved. But these doubts are excessive. Usually dissimilar theories, although differing in emphasis, yield similar results, as we noted in Chap. 2. Moreover, many theories overlap or have parallels in various ways. Thus a rights-based theory has significant differences from duty-based or rule-utilitarian theories, but the latter two still deal with conceptions of moral rights. And a rights-based theory will take seriously both duties and good consequences. There are close "family resemblances" among ethical theories which should not be disregarded in the fury of philosophical battle. In addition, theories must be assessed in the light of our judgments about practical questions raised within the professions and everyday life. As most ethicists would agree, there is a dynamic interplay between the role of systematic theories in correcting and modifying more concrete and specific judgments, and the role of those concrete judgments in testing the validity of theories (Rawls, 20 and 48).

In regard to the safety obligations of engineers, there are several stages or levels in moving from general theories to the specific actions they are used to justify. These levels may be pictured by using an architectural metaphor. Consider the tower shown in Fig. 4–13. Each of the four main stages or girders represents a type of moral claim. Girder 4 at the top has to do with claims about particular actions being right or obligatory. The next beam down symbolizes claims that specific engineers have special moral obligations or duties concerning safety. Girder 2 represents the special safety principles themselves. Candidates for inclusion here would be items appearing in engineering codes of ethics. (We say "candidates" since such principles must be established as justified in order to be included.) And the lowest stage (girder 1) is concerned with the most general and basic foundational or philosophical principles.

The vertical lines at the bottom, illustrating piles for stabilizing the foundation, represent arguments in favor of the foundational principles. Such arguments are possible, although not of course by reference to still more basic normative principles which, if they existed, would then themselves become the real foundational moral principles. Instead, the "stabilizing" arguments are concerned with such matters as a given theory's consistency, its comprehensiveness or wide applicability, and how to successfully mesh our most carefully considered moral intuitions.

The latticework connecting the girders represents the morally relevant information needed to move from one type of normative claim to another. For example, consider the move from level 3 to level 4. Suppose we agree that an engineer has an obligation to inform the proper authority of serious dangers to the public. In order to know how this obligation should be carried out correctly, then, we need to know who in the particular situation

FIGURE 4-13

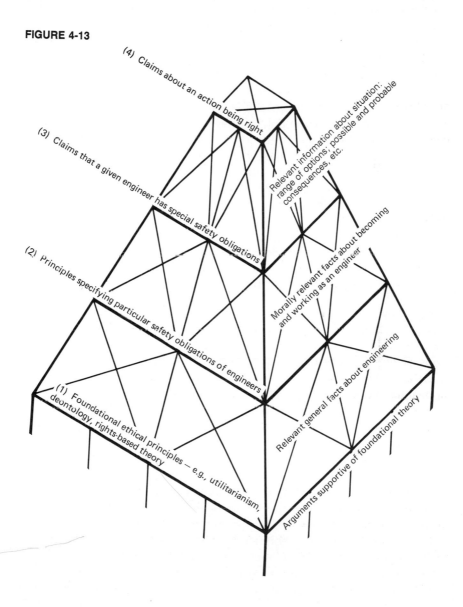

constitutes the proper authority and how that authority should be notified—by an anonymous phone call or by a formal memo delivered via certified mail?

Of main concern in the remainder of this section will be the move from stage 2 (the principles specifying particular safety obligations of engineers) to stage 3 (the claim that some given engineer has those obligations). We will also, however, sketch how a rights-based theory can be invoked at foundational level 1 to play a role in justifying claims at both levels 2 and 3.

The Basis for Special Safety Obligations of Engineers

For convenience assume that engineers' obligations concerning safety are roughly the ones given in the code of ethics of the former Engineers' Council for Professional Development. Some of the entries are fairly general: "Engineers shall hold paramount the safety, health and welfare of the public in the performance of their professional duties." Others are more specific: "Should the Engineers' professional judgment be overruled under circumstances where the safety, health, and welfare of the public are endangered, the Engineers shall inform their clients or employers of the possible consequences and notify other proper authority of the situation, as may be appropriate."

How do people acquire such obligations? It may seem that this question is purely factual and not a problem for normative ethics. People, it might be thought, acquire the safety obligations of engineers at the time they become engineers, and when this time occurs is determined by calendar dates alone.

There are two reasons why more than this needs to be said. First, it is not clear *when* people become engineers and *who* counts as engineers. Is it when they are granted their engineering degree, or when they are hired to work under the title of "engineer," whether or not they hold an engineering degree? Is it after they have interned in the profession for a certain time and then passed an examination to become certified as "registered engineers?" And does one cease being an engineer if one is reduced to doing merely technicians' tasks?

Considered alone these difficulties might be resolved by clarifying the concept of "engineer"—hence making it primarily a conceptual rather than a normative issue. But there is a second consideration, and in stating it let us focus exclusively on the central, uncontroversial, instances where we are sure we are talking about engineers: We need to know what it is about being an engineer that makes them have special safety obligations. And specifying this "what" is a question in normative ethics since it calls for an interpretation of the moral import of various facts about engineering.

The answer to the question may seem obvious. Yet we think it is worthwhile to state the obvious here. We shall give our own view after first discussing and rejecting four other views which strike us as either false or insufficient:

1 The first view is that engineers acquire moral obligations concerning safety by being subject to laws or enforced codes that require them to be so obligated. This constitutes a legalistic approach to morality, that is, an attempt to model morality on the law or reduce it to legal and paralegal considerations. When generalized to all morality, it is a version of what we referred to earlier as "ethical conventionalism": the doctrine that morality is nothing more than the dominant conventions, mores, or laws operating within a given society. And we reject this view, just as we rejected ethical conventionalism.

2 The second view is that engineers acquire special obligations by joining a professional society and thereby agreeing to live by that society's code of ethics.

This view differs greatly from the first by emphasizing the voluntary act of agreeing to abide by a code rather than passively being subject to enforced laws and codes. Like the next two views to be discussed, it places the origin of engineers' obligations to safety in a personal commitment to act according to principles implying ethical obligation. Thus it ties directly into our ordinary understanding of how promises and other self-committing acts generate obligations. Yet it is also doubly insufficient. What of the many engineers who choose not to join professional societies? Surely they are not released from the responsibility to meet obligations to ensure safety. But if the failure to join a society does not remove such obligations, then the act of joining cannot be the sole or main origin of acquiring them. Moreover, there is always the question of whether what is promised is ethical or not. Thus, if the code of a professional society happened to contain morally harmful entries—such as one restricting responsible criticism of other engineers' safety judgments—the promise to obey that code would either be nullified or overridden by other moral considerations.

3 The third view is that engineers acquire safety obligations through the contractual agreements by which they are hired by their companies or employers. Here we may agree immediately that *some* special safety obligations are acquired in this way. An engineer who is hired as a safety inspector surely does acquire special work responsibilities related to safety. And those responsibilities cannot be reduced merely to prudential concerns, i.e., reduced to the attitude that not meeting them would lead to the loss of one's job or a promotion. Rather they entail specific commitments to fulfill such job related duties, and this means *moral* obligations have been generated.

Yet even explicit obligations to employers cannot be the sole basis for the safety obligations of engineers. For no engineer is obligated by her or his employee status to sacrifice safety by following an unscrupulous employer's directives to lie, cheat, forge, or directly risk innocent lives by producing or approving shoddy designs and constructions. It is not empty rhetoric, or at least it should not be, to insist that general safety obligations to the public can override obligations to employers. Rather it reflects the point made earlier that specific promises and voluntary commitments sometimes must give way to wider obligations.

4 A fourth view is that engineers, upon entering their careers, made a broad, tacit promise to the public to protect and safeguard it in the course of performing their tasks. In return the public has largely underwritten their education through financial support for schools and implicitly granted the profession as a whole certain privileges. For example, it allows professional societies to accredit schools of engineering, and to participate in setting standards for the title of "professional engineer," as well as establishing technical standards. In principle it could also grant individual engineers the right to zealously pursue public safety, backed by the support of professional societies. An analogy to this would be how the public has granted doctors and lawyers the special privilege of keeping patient and client information confidential so as to increase the trust needed for successful medical therapy and legal defense (Freedman, 1978, 13). Because of these privileges, it is claimed, as well as the public's expectations, it would be unfair of professionals not to reciprocate by committing themselves to promoting those aspects of the public good that fall within their sphere of activity—in the case of engineers, to promoting public safety.

This amounts to the view that engineers have tacitly signed a kind of mutual contract with the public, and it is appealing, although the shared understanding needed to make sense of such a contract is currently rather limited. Nevertheless, if this idea of a contract is to be more than just a metaphor, it remains to be seen how such contracts are justified. That is, are such mutual agreements morally permissible or even obligatory? Answering this question takes us beyond the mere idea of agreements and reciprocal commitments to the issue of *justification* for whether those commitments ought to be made in the first place.

Different ethical theories will provide different accounts of that justification. Our own view comes out of a rights-based perspective, which emphasizes how such contracts with the public collectively represent and express the right of individuals to be protected. Accordingly, we will conclude this chapter with a brief sketch of what that perspective entails.

A Rights-Based Perspective on the Safety Obligations of Engineers

What is it then which obligates engineers to pay special attention to safety? Part of the answer is of course obvious. It is that in virtue of their special knowledge, skills, and functions, the work of engineers has an enormous impact on the safety of those affected by technological projects and products. But precisely why this obligates engineers not to develop harmful products or engage in harmful projects, and to what extent these obligations hold, are not obvious, as was said earlier. The issue can, however, be examined by applying a rights-based perspective according to which every person has an inherent human right to pursue his or her legitimate interests, i.e., interests not harming others (Melden, 1977).

Does this imply an unqualified right not to be harmed by technological products? No. If people purchase hanggliders and then kill themselves by flying them carelessly or under unfortunate weather conditions, their rights have not been violated—so long as advertisements about the joys of hanggliding did not contain misleading information. But the basic right does imply a right not to be poisoned, maimed, or killed by technological products whose dangers are not obvious or are deliberately hidden. This in turn implies a right to informed consent when purchasing or using products or services that might be dangerous (e.g., buying an airplane ticket). We might think of this as a right to make an "informed purchase."

As we have argued, engineers can be viewed as experimenters on a societal scale. And their experimental subjects have a right to judge the acceptability of the risks confronting them when they use or are affected by technological products or projects.

These rights place correlative obligations on those creating products or engaging in projects to acquire and transfer relevant safety information to consumers and others affected by the products or projects. The nature of such obligations, in general outline, will be shaped by the rights. Thus there is a direct link between basic human rights and the safety obligations of engineers, both in regard to what those obligations are and how they are acquired. And laws, professional codes, and employment agreements are important to that linkage insofar as they help ensure that the safety obligations are fulfilled. Hence, to refer back to our tower in Fig. 4–13, foundational moral rights appearing at the foundational level 1 will be used in justifying claims at all of the levels above.

In conclusion, we might distinguish between two different senses in which it is sometimes claimed that engineers have *special* safety obligations in regard to their work. If "special obligations" refer to obligations not grounded in the general human rights which play a central role in ordinary morality by placing obligations on *all* people, then the only special obligations of engineers are those arising out of special employment agreements or agreements with professional societies. But the main safety obligations of engineers are *not* due to some special professional-society membership, law, tradition, or employment condition inapplicable to nonengineers. If, however, the word "special" is applied to obligations to give special care and attention to safety matters concerning the projects they engage in, then all engineers do have special safety obligations. They have them in virtue of how their particular expertise and functions are directly related to the rights of persons affected by their work. In this sense we can say that an examination of the special professional obligations of engineers in regard to safety meshes straightforwardly with an examination of general human rights, and this reflects the direct link between engineering ethics and moral philosophy.

Study Questions

1 Sketch a rule-utilitarian justification of the special safety obligations of engineers listed in the National Society of Professional Engineers code (see Appendix). Then sketch a duty-based justification for those obligations. What would act-utilitarianism have to say about them?

2 An engineer visits a construction site where a structure designed by him is being erected. He has not been hired to supervise the construction. Noticing some unsafe conditions (poor scaffolding and the like), he wonders whether or not to report them. He remembers that on a previous job a colleague had early on reported some safety violations; then, when on later visits to the site she had not noticed additional safety violations which subsequently caused injuries to workers, she had been sued for carelessness. Her first reports had placed her in jeopardy! What should he do?

3 "Airless" paint spray guns do not need an external source of compressed air connected to the gun by a heavy hose (although they do need a cord to attach them to a power source) because they incorporate within themselves a small electric motor and pump. One common design uses an induction motor which does not cause sparking since it does not require a commutator and brushes (which are sources of sparking). Nevertheless the gun carries a label, warning users that electrical devices operated in paint spray environments pose special dangers. Another type of gun which, like the first, also requires only a power cord, is designed to weigh less by using a high-speed universal motor and a disk-type pump. The universal motor *does* require a commutator and brushes, which cause sparking. This second kind of spray gun carries a warning similar to that attached to the first, but it states in addition that the gun should never be used with paints which employ highly volatile and flammable thinners such as naphtha. The instruction booklet is quite detailed in its warnings.

A painter had been lent one of the latter types of spray guns and was operating it while it was partially filled with paint thinner in order to clean it out. It caught fire, and the painter was severely burned as the fire spread. The instruction booklet pointing out the spray gun's dangers was in the cardboard box in which the gun was kept, but it had not been read by the painter, who was a recent immigrant and did not read English very well. Do you see any ethical problems in continuing over-the-counter sales of this second type of spray gun? What should the manufacturer of this novel, light-weight device do?

In answering these questions, consider the fact that courts have ruled hidden design defects are not excused by mere warnings attached to the defective products or posted in sales rooms. Informed consent must rest on a more thorough understanding than can be transmitted to buyers by simple warning labels.

4 Write a book report on Martin J. Bailey's *Reducing Risks to Life: Measurement of the Benefits* (see Bibliography). Discuss the ethical aspects of the various methods of measuring benefits and costs described by Bailey.

5 Prepare a case study on the design of British airships from a safety point of view. Using the writings of, among others, H. C. Meyer and D. H. Robinson cited in the Bibliography, show how personality traits and politics may have affected the design of the R 101 and discuss the moral issues involved.

SUMMARY

A *risk* is the potential that something unwanted and harmful may occur. A thing is *safe* for persons to the extent that they judge (or would judge) its risks to be acceptable in the light of full information about the risks and in light of their settled value principles. Thus, in designing for safety, estimates must be made of which risks are acceptable to clients and the public that will be affected by the projects or products in question.

Many factors influence people's judgments when they decide which risks are acceptable. Most basic is their set of value principles regarding what they care about and to what extent they care about it. Other factors include whether or not a risk is assumed voluntarily, the degree of anxiety connected with any risk, whether or not a risk is immediately evident, whether or not potential victims are identifiable beforehand, and the manner in which statistics about a risk are presented to people.

Thus, for engineers, assessing safety is a complex matter. First, the risks connected to a project or product must be identified. This requires foreseeing both intended and unintended interactions between individuals or groups and machines or systems. Second, the purposes of the project or product must be identified and ranked in importance. Third, the costs of reducing risks must be estimated. Fourth, the costs must be weighed against both organizational goals (e.g., profit, reputation for quality, avoiding lawsuits) and degrees of acceptability of risks to clients and the public. Fifth, the project or product must be tested and then either carried out or manufactured.

Uncertainties in assessing risks arise at all these stages. For example, at the stage of testing a product only one or a few prototypes are typically used, and at that under carefully controlled conditions. Results may not accurately mirror what will happen following mass production and installation under normal operating conditions. Moreover, testing a product to destruction (the most effective way of testing) is sometimes ineffective or inappropriate. It may be only possible to work with simulations, including analytical tools such as *fault-tree analysis* (tracing possible causes of a systems failure back to the component level).

In spite of the complexities involved, a great many risks can be reduced or eliminated in fairly obvious and routine ways, at least by those engineers and managers who work with an attitude of deep caution. And increasingly the specter of legal liability serves as an incentive toward a preventive, defensive approach to engineering. Conceiving of engineering as a social experiment helps foster such an attitude. Moreover, by emphasizing the notion of informed consent, the experimentation model points out how safety is ultimately a matter of informed judgment about the acceptability of risks.

Philosophical ethical theories can be applied to establish the moral foundations underlying the special safety obligations of engineers. And while utilitarian and duty-based theories have something to say on the subject, a rights-based theory is especially compatible with a view that emphasizes the right of those affected by engineering to give their informed consent before being so affected. That "natural" right can be appealed to directly in showing why engineers have special safety obligations in their work.

PART III

ENGINEERS, MANAGEMENT, AND ORGANIZATIONS

Professionals have to have autonomy. They cannot be controlled, supervised, or directed by the client. Decisions have to be entrusted to their knowledge and judgment. But it is the foundation of their autonomy, and indeed its rationale, that they see themselves as "affected with the client's interest."

Peter F. Drucker

Mankind cannot survive without technology. But unless technology becomes a true servant of man, the survival of mankind is in jeopardy. And if technology is to be the servant, then the engineer's paramount loyalty must be to society.

Victor Paschkis

CHAPTER 5
PROFESSIONAL RESPONSIBILITY AND EMPLOYER AUTHORITY

The 1970 Clean Air Act requires car manufacturers to conduct 50,000-mile durability tests on new engines using only one tuneup. Test results on emissions must be reported to the Environmental Protection Agency (EPA), which decides whether the engines meet current pollution standards. In May 1972, top managers at Ford Motor Company were eagerly awaiting government approval of the test results they had submitted on the engines for 1973 Ford cars. They had every reason to be confident of the results they had submitted to EPA, which were based upon tests conducted by their own employees; their only concern was about meeting tight production schedules once EPA's approval was received.

Their confidence was shattered, however, when then Ford President Lee Iacocca received a memo from a specialist in the computer division. That computer specialist had been examining the computer tapes from the tests as part of a review of the effectiveness of his division in supporting the engine and foundry division which had developed and tested the new engines. His memo identified numerous irregularities in the test records showing unauthorized maintenance of which EPA was not notified. The memo also stated that when the specialist sought an explanation about the irregularities from the engine division he was urged to burn the computer tapes and forget the matter.

Intensive research into the matter by management quickly verified the information contained in the memo. Evidently, four "supervisory technical" employees who had conducted the original tests had ordered or engaged in over 300 acts of illegal maintenance on the test engines. Spark plugs and points had been replaced frequently, carburetors cleaned, and ignition timing repeatedly reset. These adjustments lowered the levels of pollutants emitted.

Within 3 days Mr. Iacocca revealed to EPA officials all he had learned about the tests and withdrew Ford's application for certification of four major types of engines. In spite of its full cooperation with EPA investigators, the company was fined $7 million in criminal and civil fines for having conducted improper tests and issued false government reports. Because of the record size of the fines, Ford received damaging publicity from front-page newspaper articles (for example, in the *New York Times* and the *Los Angeles Times,* February 14, 1973). It was also hurt by the costs of new tests that had to be conducted on an around-the clock emergency basis and by having to delay production schedules (*Wall Street Journal,* May 25 and 31, 1972).

Misguided Loyalty?

Nothing written about the Ford test scandal tells what motivated the Ford supervisors and other engineers and technicians involved. Possibly it was only a self-interested concern to make themselves look good by ensuring their engines would pass the qualifying tests. But it is equally possible that they were acting as loyal employees. Ford had been late in obtaining some government approvals the previous year, and perhaps the individuals believed—however mistakenly—that they were serving the company's best interests by avoiding such difficulties this year. Perhaps some of them were merely following orders from higher up to tamper with the engines. In any case, management was not particularly punitive about what happened since, despite the staggering costs incurred, no one who had participated in rigging the tests was fired and the four supervisors were merely transferred to new positions.

This case suggests three points concerning the relationship between professional responsibility and loyalty to companies or employers. First, acting on professional commitments to the public can be a more effective way to serve a company than a mere willingness to do anything one sees as good for the company. Ford would have benefited much more from engineers committed to professional standards than it did by the misguided loyalty shown to it by its employees.

Second, it is clear from the example that loyalty to companies or their current owners should not be equated with merely obeying one's immediate supervisor. It would have shown a greater loyalty to Ford to act in a way consistent with the concerns of higher management, rather than in a manner consistent with the aims of an immediate supervisor.

Third, the case illustrates how an engineer might have professional obligations to both an employer and to the public that reinforce rather than contradict each other. Thus there need be *no general contrast* between the status of employee and the status of professional. In fact, obligations to the public and to one's employer often point in the same direction.

Nevertheless, we have also seen from other cases we have examined that obligations to employers and to the public do not always straightforwardly coincide. Sometimes an engineer seeking to protect the public is overruled by top management for financial reasons. For example, in the DC-10 case, the director of product engineering was told by higher management that it would be too costly for his company to redesign an unsafe cabin floor and cargo door. At other times there are disagreements over technical matters, and engineers are told they must not push their own views further. This we saw illustrated in the BART case.

The relationship between being a responsible engineer, having obligations to the public, and being a loyal employee is a matter of some complexity. We will explore it first from the direction of professionalism, then from a study of employers' authority, and finally by discussing the topics of conflicts of interest, confidentiality, and unionism.

PROFESSIONALISM

What is a professional? If we answer that it is someone who is a member of a profession, then what is a profession and how does one become a member of one? Our first concern in dealing with these questions is to understand why there is so much disagreement over how to answer them. A second concern is to sketch a conception of professionalism compatible with viewing employed engineers as professionals having obligations to both employers and the public.

Professions

In one of its senses, the word "profession" is used as a synonym for "job" or "occupation," and to be a professional at some activity means merely to earn one's living through it. Thus we speak of professional football and tennis players, as opposed to amateurs who do not draw an income from these sports. We also speak of professional sanitation workers, taxicab drivers, bartenders, and even mercenaries and killers.

But there is another sense of the word which rules out such examples. According to that understanding, professions are only certain occupations which meet special criteria. Generally the criteria include restrictions of the following sort:

1 The work involves exercising sophisticated skills, judgment, and discretion which is not entirely routine or susceptible to mechanization.

2 Preparation to engage in the work requires extensive formal education, including technical studies in one or more areas of systematic knowledge as well as broader humanistic studies. Generally, continuing education and updating of knowledge are also required.

3 Special societies and organizations controlled by members of the profession are allowed by the public to play a major role in setting standards for admission to the profession, drafting codes of ethics, enforcing standards of conduct, and representing the profession before the public and the government.

4 The occupation serves some important aspect of the public good, as indicated in the codes of ethics. (For example, medicine is directed toward promoting health, law toward protecting the public's legal rights, and engineering toward promoting the public's health, safety, and welfare as they relate to technology.)

There are many debates over just which occupations meet these criteria. The traditional professions of medicine, law, teaching, and the ministry are cited as paradigm or clear-cut examples. So too are professions like engineering and business administration that have emerged more recently. Sanitation work, taxicab driving, and basketball are not counted because of the lack of required advanced education. Disagreements occur over occupations requiring intermediate amounts of formal training: advertising, realty, cosmetology, and some jobs in computer and medical technology.

Membership Criteria

Further disputes arise over how a person does or should become a member of an accepted profession. Such disputes often occur with respect to engineering. Each of the following has been proposed as a criterion for being an engineer or a "professional engineer" in the United States:

1 Earning a bachelor's degree in engineering at a school approved by the Accreditation Board for Engineering and Technology. (If applied in retrospect, this would rule out Leonardo da Vinci and Thomas Edison.)

2 Performing work commonly recognized as what engineers do. (This rules out many engineers who have become full-time managers, and also rules in some people who do not hold engineering degrees.)

3 Being officially registered and licensed as a "Professional Engineer" ("P.E."). Becoming registered typically includes (a) passing the Engineer-in-Training Examination or Professional Engineer Associate Examination during the senior year in engineering school, (b) working 4 to 5 years at responsible engineering, (c) passing a professional examination, (d) paying the requisite registration fees. (This rules out a large percentage of unregistered people holding bachelor's, master's, and doctorate degrees in engineering, many of whom work in education or manufacturing industries where they are exempt from registration.)

4 Acting in morally responsible ways within engineering. (This rules out scoundrels, no matter how creative they may be within engineering.)

The words "profession" and "professional" have acquired positive emotional connotations and suggest a highly desirable status for occupations and individuals. At least part of these connotations derive from the public importance of professional skills and knowledge, and also from the difficulty in acquiring them. Because of these factors, professionals are regarded as deserving high pay, prestige, and other social benefits. Social status is frequently enhanced by a title, such as "Doctor" or "Reverend." In this respect some engineers in the United States, where engineering is often not considered on a par with medicine or the ministry, yearn for more of the open recognition accorded their counterparts in some other countries.

Persuasive Definitions

One could choose any one of the above-mentioned criteria as to what constitutes an engineer and claim, by assuming a particular value perspective, that it is the only correct definition. The somewhat loose ordinary cognitive meaning (defining criteria) could then be altered by making it more precise and narrow while retaining the ordinary *emotive meaning* (positive connotations). One would then be giving what is called a *persuasive definition* of the term "professional engineer": one used to espouse a particular value perspective (Stevenson; Cogan, 105).

As might be expected, such persuasive definitions occur frequently in disagreements over values, and there need be nothing disreputable about them. But they must be understood for what they are: techniques for altering attitudes which by themselves do not constitute arguments. They should be critically examined, rather than passively allowed to influence us under the guise of being "truths, by definition." For they are not at all like definitions of triangles as three-sided planar figures or bachelors as adult, unmarried males.

For instance, if a psychologist defines intelligence as simply what certain psychological tests reveal, we should beware of the possible implications of bestowing so much significance on present-day techniques of psychological testing. Again, if medicine is defined as the science of health, and if health is defined as "a state of complete physical, mental and social well-being" (which is how the World Health Organization defines it), we should be wary of how these definitions encourage excessive expectations about what doctors and medical techniques can do.

Similarly, we need to be ready to assess the implications of accepting any given persuasive definition of professional engineer. The attitudes and value perspectives embodied in such definitions concerning the desirable properties of professional engineers and how best to identify those properties need to be critically examined. For instance, those who seek to restrict the term "professional" to officially registered engineers will view the restriction as a way to ensure that stringent qualifications for engineering are met which will maximize benefits to the public. Those who are against this definition, however, may argue that it needlessly increases bureaucracy and is not an effective way of judging engineering qualifications.

Professionalism as Independence

There is one type of persuasive definition of professional engineer which is especially significant for our present purposes. That definition directly ties professionalism to independence and freedom from coercion. One version was given by Robert L. Whitelaw in an essay entitled, "The Professional Status of the American Engineer: A Bill of Rights." Whitelaw sharply contrasts bureaucratic submission to employers with the independence he sees as inherent in professionalism. In fact, he defines professionalism and employee status as logically incompatible: ". . . so long as the individual is looked upon as an employee rather than as a free artisan, to that extent there is no professional status" (Whitelaw, 37–38).

In Whitelaw's view, only consulting engineers qualify as professionals. The mass of engineers working as employees within corporate or governmental bureaucracies will not become professionals until they are protected by an engineering bill of rights ensuring the freedoms already enjoyed by self-employed engineers. Examples of these rights are "the right to refuse unethical activity without prejudice or loss of contract" and "the right to freedom from surveillance, psychological manipulation, and other job evaluation techniques."

According to Whitelaw's definition, one is not a professional engineer if one acts merely on the basis of an employer's orders in matters where the public good is concerned. Being a professional involves the freedom to act according to one's own judgment about what the correct course of any action should be. It is clear that Whitelaw is reacting sharply to what he views as the overdomination of engineers by the authority of management. While many of his concerns are legitimate, and while his definition is a potent rhetorical instrument, we must ask whether his definition expresses too extreme a position.

Professionalism as Serving Employers

An opposite type of persuasive definition would treat loyal service to employers (or to clients, in the case of consulting engineers) as the heart of professionalism in engineering. Such a view is implied in Samuel Florman's widely discussed essay, "Moral Blueprints." Florman argues that "it is essential that professionals should serve" (Florman, 1978, 32). Rather than "filtering their everyday work through a sieve of ethical sensitivity," as Florman puts it, professionals have the task of meeting the expectations of their clients and employers. Professional restraints should be laws and government regulations rather than personal conscience.

Florman's essay is devoted to attacking the entry in the code of ethics of the former Engineers' Council for Professional Development which states, "Engineers shall hold paramount the safety, health and welfare of the public in the performance of their professional duties." His response is: "Engineers are obliged to bring integrity and competence to whatever work they undertake. But they should not be counted upon to consider paramount the welfare of the human race" (Florman, 1978, 32).

It is fair to view Florman as expressing the dominant view of management concerning engineering professionalism. And many engineers would concur with the definition. Yet here again we must ask whether this conception of the professional obligations of engineers is one-sided.

An Intermediate Position

We will state, but not attempt to defend in detail, our own "persuasive definition" of professionalism in engineering. Our main concern in this section has been to emphasize that such definitions will generally be an outgrowth of one's perspective on the moral obligations of engineers. Accordingly, in discussions about the subject, attention should be focused on the obligations themselves rather than on how they are reflected in the criteria one espouses in defining the term "professional engineer."

Our view of the obligations of engineers involves a moderate position lying between the extremes represented by Florman and Whitelaw. For us, employed engineers have major moral obligations to both employers and to the public, and we think it a mistake to seize on either obligation as the essence of professionalism. A more useful definition would allow us to speak straightforwardly of "salaried professionals" (contra Whitelaw), and would also enable us to reject the view of professional obligations as essentially service to employers within the law (contra Florman).

Accordingly, we favor viewing professional engineers as meeting two general criteria: (1) Attaining standards of achievement in either education, job performance, or creativity in engineering which distinguish them from engineering technicians and technologists. (We recognize that for legal and educational purposes the nature of those standards will have to be made more clear-cut and explicit.) (2) Accepting as part of their professional obligations at least the most basic moral responsibilities to the public as well as to their employers, clients, colleagues, and subordinates. This latter criterion lends to the term "professionalism" a moral dimension consistent with the fact that "unprofessional conduct" is often used as a synonym for "unethical conduct." Yet it makes no assumptions about which type of obligation is most central to engineering—an issue that should be debated independently of how to define what it means to be a professional engineer.

Obligations to the Public as Paramount

At this point let us set aside the issues over how to define professionalism and turn directly to the relationship between the two general obligations to the public and to employers. Should we agree that the obligation to protect the public health, safety, and welfare is *paramount,* as recent codes have stated?

In our view, yes, so long as "paramount" is understood in its colloquial sense to mean "chief in importance or deserving primary emphasis." We make this judgment against the background of a conspicuous reality: Most employers have enormous power compared with the engineers they employ. They have the power to fire or take other negative sanctions against individuals who fail to meet their obligations to the employer. And engineers have relatively little recourse at present to respond to an employer who does not support their efforts to meet their obligations to the public. Hence if "paramount" means "deserving most emphasis in the minds of engineers, engineering societies, and the wider community," then the obligation to the public deserves to be regarded as paramount.

"Paramount," however, can also be construed in a technical philosophical sense to mean that whenever the obligations to employers and the public come into conflict (creating a moral dilemma), the obligation to the public *always* takes precedence. Thus it can mean, to use W. D. Ross's language, that whenever these two prima facie duties conflict, one's actual duty— what one *ought* to do, all things considered—is always to meet one's public obligations.

We doubt that this technical sense of "paramount" is what drafters of the codes had in mind. In any case, it seems to us to be a dubious view if carried to its extreme. Consider the following case: A design group develops a new electronic circuit to be used in clock radios which would extend their average life from 5 to 7 years at a cost that would raise manufacturing expenses by only 1 percent. After presenting their arguments to top management, however, the latter reject the proposal as not being cost-effective. Does the design group's obligation to the public outweigh its employer's directives to drop further work on the circuit?

In this case it would undermine an employer's legitimate authority to say that engineers must subsume their obligations as employees to their obligations to the public. Of course the obligation to the public should override the obligation to the employer in cases where something of extreme importance is at stake for the public: generally where lives are seriously threatened, serious financial corruption is involved, or serious economic loss might result. Many instances of justified whistleblowing fall into this category, as we shall argue in Chap. 6. But this does not mean that the first priority is *always* the public good whenever that good conflicts morally with an employer's good.

Engineers, in short, must weigh their obligations to the public, their employers, their colleagues, and others when conflicts among such obligations arise. A simple, exceptionless ordering of priorities among obligations is not always possible.

Study Questions

1 Comment on the following definitions, or partial definitions, of professionalism in engineering. In each case, do you agree that the passage presents something essential to an understanding of professionalism? Is the definition a controversial persuasive definition with which you disagree? Why?

a "Professionalism implies a certain set of attitudes. A professional analyzes problems from a base of knowledge in a specific area, in a manner which is objective and independent of self-interest and directed toward the best interests of his client. In fact, the professional's task is to know what is best for his client even if his client does not know himself" (Storch, 38).

b "A truly professional man will go beyond the call to duty. He will assume his just share of the responsibility to use his special knowledge to make his community, his state, and his nation a better place in which to live. He will give freely of his time, his energy, and his worldly goods to assist his fellow man and promote the welfare of his community. He will assume his full share of civic responsibility" (Simrall, 39).

c "If they mean to be professionals, engineers themselves will have to take moral responsibility for their work rather than unquestioningly accepting whatever orders come down to them from Government or employers" (Walters, 1973, 42).

d "A profession, in contradistinction to a trade . . . [is] a body of persons with learned knowledge having an ability to examine itself and its purposes; an ability to link its body of knowledge with other bodies of knowledge to achieve common purposes; the ability to defend dissent, not just within the society but dissent by its members in conflicts with their employer organizations or their government agencies or corporations; and above all the ability to pioneer new policies that are not brought into effect by market incentives" (Nader, 1972, 14).

2 Discuss under what circumstances you think engineers are justified in participating in the design and manufacture of products with built-in obsolescence. Such products wear out rapidly and cannot be repaired.

EMPLOYERS' AUTHORITY

Salaried engineers have obligations to respect their employers' legitimate authority. But what is the nature of this authority? How far should it be recognized by salaried professionals as being morally justified?

In order to address these questions we will begin with a discussion of how and why authority arises within institutions. Then several distinctions will be drawn which make clear why such authority is not automatically the same as moral authority.

Goals of Institutions

Engineers work within virtually all forms of modern organizations. These organizations vary enormously in the specific goals they are created and maintained to serve. Two general types are (1) service organizations, and (2) business or profit-making organizations (Drucker, 131).

Service institutions have as their primary purpose to provide selected services to the public or to other organizations of which they are parts. Universities provide education, hospitals give health care, court systems serve legal needs, professional societies serve professionals, and so on for churches, the military and government, and "natural monopolies" like telephone and utility companies. They operate under the economic restraints of a budget allocated by supervising government agencies or based on their income from the services they provide.

Business or profit-making institutions are established primarily to produce income. The criterion for performance is taken by some to be maximum profits (Friedman, 1979, 192), and by others to be a reasonable return on investment. Still other observers will include making social contributions as well. As the necessary means to achieving the primary purpose of producing income, business institutions must of course provide some product or service which customers will purchase. Moreover, businesses must do so within the boundaries delimiting the public good set by the government, which grants them the charters that allow them to operate. Thus a fuller specification of the purpose of profit-making institutions is to make a profit by providing a product or service which the public finds useful.

Both service and business organizations may take on further secondary goals. In order to meet their primary goals, they generally adopt the goals of survival and of maintaining adequate degrees of freedom from outside control (Galbraith, 170). In practice this latter goal often means resisting extensive government regulation.

Institutional Authority

In order to meet their institutional goals, organizational rules are created. Typically these rules attach specific duties to positions within the organization. The rules may also allow one person to assign duties to others. Thus, an *institutional duty* is any assigned task within an organization, whether the assignment is directly or indirectly rule-specified. Managerial tasks, for example, may be to allocate money or other resources, to make policy decisions or recommendations, or to oversee projects and issue directives to subordinates on particular topics.

The need for authority relationships in meeting organizational goals is clear. Decisions must be made in situations where allowing everyone to exercise unrestrained individual discretion would create chaos. Moreover, clear lines of authority provide a means for identifying areas of personal responsibility and accountability.

In order to enable people holding managerial positions to meet their institutional duties, the rules also assign them the requisite authority. This is *institutional authority,* since it is acquired, exercised, and defined within institutions. It may be defined as the *institutional right* given to a person to exercise power based on the resources of the institution (Pichler, 428).

Institutional rights (authority) and duties are for the most part two sides of the same coin, and they deal with precisely the same activities and functions. Project engineers, for example, have the institutional duty to ensure that the projects they supervise are successfully completed, and they are given the institutional rights or authority necessary to carry out this duty. Obviously, too, these rights involve a certain amount of freedom or liberty: It would be self-defeating for an institution to assign tasks but to deny the freedom from interference necessary to perform them.

Institutional versus Expert Authority

It clearly benefits institutions to give authority to the individuals best qualified to serve the institution's goals in a given capacity. But in practice there is not always a perfect match between the authority granted and the qualifications needed to exercise it. Incompetence is found in all large institutions, and there is some truth in the cynical remark that in bureaucracies people tend to rise to their own level of incompetence.

Thus institutional authority should not be equated with *expert authority*.

Expert authority is the possession of special knowledge, skill, or competence to perform some task or to give sound advice. In this sense, doctors are authorities on health and civil engineers are authorities on structures and transportation. One of the key competencies for management is leadership ability, which has its own kind of expert authority that has been called the "authority of leadership": the expertise to effectively direct others (Barnard, 173).

It is possible for engineers to have expert authority in matters for which they have little or no institutional authority to make decisions. Their institutional authority may extend no further than the right to provide management with analyses of possible ways to perform some technical task, after which they are restricted to following management's directives about which option to pursue.

Authority versus Power

Institutional authority must also be distinguished from power. Institutional authority typically carries with it an allotment of the resources needed to complete tasks. Yet ineffectual persons may not be able to summon the power which their position allows them to exercise. A manager, for example, who lacks the skills of leadership may be unable to inspire and encourage employees to produce in ways the institution requires, much in the way a conductor may fail with an orchestra.

Conversely, people who are especially effective may acquire great power or influence which goes well beyond the authority attached to the positions they hold. Charismatic leaders often have influence outside their domains of authority. And highly respected engineers of proven integrity may have power within an organization exceeding their explicit institutional rights.

Authority and Managerial Strategies

As we noted, institutional authority often gives one the prerogative to issue orders in a given area and to expect them to be complied with. But it would be a serious misconception to equate managing people with issuing orders and then standing on one's authority in demanding unquestioning obedience (McGregor).

To manage people is to guide and integrate their work, and there are many general approaches or strategies for doing so. One is the direct assertion of authority over a subordinate: "I'm in charge—obey or I'll fire you." But repeated use of this approach would be viewed negatively within business as an authoritarian abuse of authority. Other strategies include a heavy mixture of persuasion and rational argument. Some emphasize mutual decision making, or decision making based on full consultation with subordinates.

A consensus approach may be slower, but it is more effective and prudent in the long run. And in dealing with salaried professionals, it is more than prudent. A strictly authoritarian approach can easily lead to a demise of moral integrity among employees with a resultant weakening of felt obligations vis-à-vis both employer and the public.

Morally Justified Authority

The preceding distinctions clear the way for making two observations. First, an employer may have the institutional authority to direct engineers to do something which is not morally justified. Second, engineers may have an institutional duty to obey a directive which is morally unjustified and which it is their moral duty, all things considered, to disobey.

To repeat: Institutional authority is the institutional right to exercise certain kinds of power, and this right is merely the liberty which the rules of the institution say a person has. Institutional duties are the duties specified by the rules of the institution, either directly as attached to offices and positions or indirectly as delegated by a superior (who in turn derives such authority from the rules of the institution). These rights and duties may be established as means to the end of meeting institutional goals. But they are not thereby moral rights and duties, or morally justified institutional rights and duties.

Before concluding that a specific act of exercising institutional authority is morally justified, we would need to know (1) whether the institutional goals are themselves morally permissible or morally desirable and (2) whether that act violates basic moral duties.

Engineers do take on some moral obligations to meet their institutional duties when they accept employment—but only so long as meeting those institutional duties is morally permissible. An employment contract can be viewed as a morally conditioned mutual promise. Promises to act immorally are either invalid or automatically overridden by moral considerations.

The relationship between *moral* rights and duties and *institutional* rights and duties is complex. Only a few further observations will be made here. Recall that in earlier chapters we distinguished between general human moral rights and special moral rights. Obviously human rights and institutional rights cannot be equated. By definition, human rights (such as the rights to life, liberty, and the pursuit of happiness) are possessed by virtue of being a person, not by virtue of being a member of an institution.

However, some institutional rights and duties can be equated with special moral rights and duties—namely, those which are morally justified. For example, through employment agreements employees acquire a special institutional duty to protect proprietary information, and employers have an institutional right to require that employees do so. And to the extent that those duties and rights can be morally justified, either through some argument deriving from the employment contract itself or because of other, independent considerations, they are also moral duties and rights.

Accepting and Obeying Authority

Let us now shift perspective from the authority of employers to the recognition of that authority by their employees. Employees recognize their employer's authority when for the most part they accept the guidance and obey the directives issued by the employer having to do with the areas of activity covered by the employer's institutional authority. There are exceptions, since it is possible in special cases to recognize someone's authority but to disobey an order on moral grounds. But our present concern is to obtain a clearer idea of what accepting authority under normal conditions should and should not involve.

In his classic text, *Administrative Behavior,* Herbert Simon states: "A subordinate is said to accept authority whenever he permits his behavior to be guided by the decision of a superior, without independently examining the merits of that decision" (Simon, 11). In general, authority relationships are "all situations where suggestions are accepted without any critical review or consideration" (Simon, 128). Again, "the characteristic which distinguishes authority from other kinds of influence is . . . that a subordinate holds in abeyance his own critical faculties for choosing between alternatives and uses the formal criterion of the receipt of a command or signal as his basis for choice" (Simon, 127). In Simon's view, the reasoning of subordinates in their role as subordinates is at most aimed at anticipating commands by asking themselves how their superiors would wish them to behave in a given situation.

Simon notes that all employees place limits on the "zone of acceptance" in which they are willing to accept their employer's authority. But within that zone, an "individual, relaxing his own critical faculties, permits the communicated decision of another person to guide his own choice" (Simon, 151).

Simon provides an influential picture of what obedience involves. But its limitations must be kept in mind. Employees are generally not inclined to make an issue of every incident of questionable morality, sometimes because of moral inertia, at other times out of a reluctance to generate an overload of complaints or a willingness to give their employer a certain amount of leeway within which to operate, or even because of a wish to save the strongest arguments and the possible risk of losing their job for the most serious infractions. While this automatic obedience within the "zone of acceptance" of an employer's authority is understandable, it also carries with it the risk of becoming blind and unthinking in regard to moral matters. The problem which arises then is that the boundaries of tolerance are easily expanded and rationalized when expediency so dictates. Thus the size of any person's "zone of acceptance" could become a measure of the lack of that individual's moral integrity. To avoid this problem, employees must be reflective concerning the justified extent of their "zone of acceptance" of employers' authority. In a sense, then, they should never suspend their critical review of employers' directives in the manner Simon describes.

From a different direction, therefore, we have reached the same conclusion we came to in the last section: As professionals, engineers have obligations to accept their employers' institutional authority. But this is not an obligation to obey blindly. Professional autonomy entails exercising independent judgment, even though it does not mean disregarding legitimate directives. The basic moral task of salaried engineers is to be aware of their obligations to obey employers on the one hand and to protect and serve the public and clients on the other. Most of the time there is no conflict between the two. But when occasionally genuine conflict arises, it must be resolved by the exercise of an autonomous moral judgment.

Loyalty

Let us return for a moment to the topic of loyalty to company and employer, a topic mentioned in connection with the Ford case at the beginning of this chapter. The word "loyalty" suggests something more than merely recognizing and accepting the authority of the employer. It implies, at least in ordinary language, doing so from certain kinds of motives. People who detest their employers and companies and who obey grudgingly and spitefully are not considered loyal. A loyal person shows at least some degree of genuine concern in serving the interests of those to whom she or he is loyal.

Actually there are two different concepts of loyalty. According to the first, to be loyal and faithful is to seek to meet one's moral duties to a person or organization, and to do so willingly or even with zealous devotion (Ladd, 1967, 98). In this sense loyalty is an inherently good thing. Indeed, it is a moral virtue.

According to the second concept, by contrast, loyalty is not automatically a good thing. To be loyal and faithful means to be devoted and obedient to or zealously supportive of a cause, person, or organization, but not necessarily out of moral duty. People loyal in this sense try to promote the interests of whatever or whomever they are loyal to and they do so out of genuine concern. But whether it is good or obligatory to be loyal in this way depends upon the specific person, organization, or cause the loyalty is directed toward, and upon the circumstances in which the loyalty is displayed. There is a moral obligation *not* to act loyally in situations where violations of important moral duties could occur.

When codes of ethics state that engineers ought to be loyal to employers, or that they should act as their employer's or client's "faithful agents or trustees," the word is generally meant in its moral sense, as is suggested by the subheadings under the injunction to be a faithful agent or trustee. Typically those subheadings list specific moral duties: to avoid conflicts of interests, to inform employers of any possible conflicts of interests, to protect confidential information, to be honest in making estimates, to admit one's errors, and so on.

Yet it is important to bear in mind the possible ambiguity in speaking of loyalty. A call for loyalty to a company may be intended as more than a call for meeting one's moral obligations, and may involve the second concept of loyalty. It can be a tacit urging of close emotional identification with, and personal commitment to, the company's good. Urging loyalty to an employer can even mean recommending unquestioning obedience and devotion to the employer.

Loyalty and faithfulness in this second sense can be very valuable in creating a climate of mutual concern and commitment to shared goals among members of an organization. It can add a human and personal dimension to the workplace, as well as aid in meeting the organization's goals. Yet it also has the potential in some situations of leading employees to disregard their wider moral obligations. For it can encourage the uncritical attitude that whatever is good for one's company is automatically good for the public.

Study Questions

1 "An applicant for employment in a number of companies accepted employment with Company X, knowing that he preferred employment in Company Y. He did not get an offer from Company Y until after he had worked for Company X for three months. He then changed to Company Y, and after several months there he discovered that employment conditions were not as good as they were in Company X. He then applied at Company X for re-employment" (Alger, Christensen, and Olmsted, 219).

In the preceding extract, did the person fail to act loyally to Company Y? In answering this question, distinguish and discuss both concepts of loyalty mentioned in this section. Touch also upon the element of time of service as it may relate to loyalty.

2 The moral complexities of obeying authority arise in most contexts where authority is needed for meeting specific goals of a group. In this connection, discuss any analogies or dissimilarities you see between the obligations of employed professionals to obey employers and accept their authority and (a) professional baseball players obeying umpires, especially in cases where the umpire makes a bad call; (b) children respecting their parents' authority; (c) soldiers on a battlefield obeying their commanders; (d) college students recognizing their professor's authority to direct a class; (e) nurses obeying doctors' orders and the directives of hospital administrators; (f) musicians obeying a conductor. Which of these contexts has the closest analogies to why employees generally ought, and perhaps occasionally ought not, to obey their employers? In presenting your answer consider some examples where those in authority make an incorrect decision or issue a poor directive.

3 Consult a recent management text for its descriptions of management strategies. Drawing upon those descriptions, distinguish and evaluate the relative effectiveness of authoritarian and nonauthoritarian approaches.

4 How can the concept of employees' loyalty to employers be upheld in the case of a company which falls into one or more of the following categories: (a) rapidly expands its work force—including engineers—when its business is good, but equally rapidly lays off employees when business begins to drop; (b) is bought out by a conglomerate with headquarters in a distant city and with more apparent interest in the acquired company's profit-making potential than its products; (c) is owned by shareholders who buy or sell shares at a moment's notice, depending on the daily stock market report.

CONFLICTS OF INTEREST

Engineers are expected to avoid conflicts of interest and to protect confidential information. Traditionally these two obligations have been given prominence in engineering codes of ethics, in management policy statements, and in the law. Indeed, next to following legitimate directives, they are probably the most emphasized aspects of displaying loyalty to employers and companies. This section will focus on conflicts of interests, and the next section on confidentiality.

Definition

In a wide sense, conflicts of interest arise whenever people or groups have interests which if pursued could keep them from meeting at least one of their obligations. We are concerned with the obligations of employees to serve the interests of their employers or companies. Thus we will mean by "employee conflicts of interest" any situation where employees have an interest which if pursued might keep them from meeting their obligations to serve the interests of their employers or companies.

Sometimes such an interest involves serving in some other professional role—say, as a consultant for a competitor's company. Other times it is a more personal interest, such as making substantial private investments in a competitor's company.

These side interests are generally understood to threaten the fulfillment of employer-related obligations in one main way: They have the potential to deflect or distort the judgment of at least some people who find themselves in that type of situation. Thus an alternative definition of employee conflicts of interest is the following: Situations where employees have side interests substantial enough potentially to affect their independent judgment, or the independent judgment of a typical person in their situation, in serving their company's interests. The qualification concerning "a typical person" is necessary. There might be conclusive evidence that the actual people involved would never allow a side interest to affect their judgment. But they could still be said to be in a conflict of interest situation.

Being in such a situation is not merely being confronted with conflicting interests (Margolis, 361). A student, for example, may have interests in excelling on four final exams. She believes, however, that there is time to study adequately for only three of them, and so she must choose which interest not to pursue. Or an investor may strongly desire to invest in two stocks but have sufficient funds for investment in only one. In these cases "conflicting interests" means a person has two or more desires which cannot all be satisfied given the circumstances. But there is no suggestion that it is *morally* wrong or problematic to try pursuing them all. By contrast, in conflicts of interest it is often physically or economically possible to pursue all of the conflicting interests, but it is morally problematic whether one should do so.

Conflicts of interest should also be distinguished from moral dilemmas, even though in some situations both are involved. Moral dilemmas occur when two or more moral obligations, rights, or ideals come into conflict and not all of them can be met. By contrast, it is often possible for an employee caught in a conflict of interest to pursue both the obligation to the employer and the side interest.

Examples

A wide variety of circumstances might arise which create conflicts of interest for employees. One type already mentioned is having an interest in a competitor's business. This might involve actually working for the competitor as an employee or consultant. Or it might involve partial ownership or substantial stockholdings in the competitor's business.

A variation on this type of situation is when engineers prepare to leave a corporation to form their own competing businesses. It would be a clear conflict of interest if they sought to lure customers away from their current employers while still working for them.

A second important category involves using "inside" information to gain an advantage or set up a business opportunity for oneself, one's family, or one's friends. Thus, for example, engineers might tell their friends about their corporation's plans for a merger which will greatly improve the worth of another company's stock. In doing so, they give those friends an edge on an investment promising high returns. In general, the use of any company secrets by employees to secure a personal gain is felt to threaten the interests of the company the employees are supposed to serve and thus to constitute a conflict of interest.

A third typical variety arises when employees benefit from personal involvements with suppliers, subcontractors, or customers. An obvious example is accepting bribes directly intended to influence judgment. Bribes may be in the form of cash, gifts, loans, services, trips, or entertainment. Another blatant example is working for a subcontractor or supplier who deals with one's corporation. Conflicts of interest also arise by holding substantial stock or other investments in a firm with which one's company does business.

Sometimes it is difficult to determine just when conflicts of interest exist. Does holding a few shares of stock in a company one has occasional dealings with constitute a conflict of interest? How about occasional luncheons paid for by vendors giving sales presentations? Or those free pens and bottles of wine from salespersons? What about a gift one believes is based on friendship rather than intended to influence one's judgment?

The code of ethics of the former Engineers' Council for Professional Development seems to recommend a hard line on such gratuities: "Engineers shall not solicit nor accept gratuities, directly or indirectly, from contractors, their agents, or other parties dealing with their agents, or other parties dealing with their clients or employers in connection with work for which they are responsible" (Sec. 4-e).

Yet most employers would consider this position excessive. Company policies generally ban any gratuities which have more than "nominal" value, or which have any realistic potential for biasing judgment. In part the specific criteria for "nominal" value will be what is widely and openly accepted as normal business practice. In part it will be assessed by a person's own awareness of what might influence his or her judgment. And in part it must be weighed according to how others might perceive (or misperceive) the gratuities. Companies also typically formulate policies about what is a nonnominal gift from a salesperson: for example, more than a $10 item, or items totaling over $30 per year.

Moral Status of Conflicts of Interest

There are many other kinds of conflicts of interest we could mention. For example, there is taking additional outside employment—moonlighting—in situations where it harms on-job performance (Reed, 19–23). And there are cases involving confidential information, to be discussed in the next section. Conflicts of interest can arise in innumerable ways, and with many degrees of subtlety. But let us now ask the following: What is wrong with employees having conflicts of interest?

Most of the answer is obvious from our definition: Employee conflicts of interest occur when employees have interests which if pursued could keep them from meeting their obligations to serve the interests of the company for which they work. Such conflicts of interest should be avoided because they threaten to prevent fully meeting those obligations.

More than this, however, needs to be said. Why should mere threats of possible harm always be condemned? Suppose that substantial good might sometimes result from pursuing a conflict of interest?

In fact it is not always unethical to pursue conflicts of interest! In practice some conflicts are unavoidable, or even acceptable. One illustration of this is how the government allows employees of aircraft manufacturers, like Lockheed or McDonnell Douglas, to serve as government inspectors for the Federal Aviation Agency (FAA). The FAA is charged with regulating airplane manufacturers and making objective safety and quality inspections of the airplanes they build. Naturally the two roles of government inspector and employee for the manufacturer being inspected could lead to a conflict of interest and biased judgments. Yet with careful screening of inspectors, the likelihood of such bias is said to be outweighed by the practical necessities of airplane inspection. The options would be to greatly increase the number of nonindustry government workers (at great expense to taxpayers) or to do without government inspection altogether (at the risk of public safety).

Where conflicts of interest are unavoidable or reasonable, employees are still obligated to inform their employers and obtain an approval. This suggests a fuller answer to why conflicts of interest are generally prohibited: The professional obligation to employers is (1) very important in that it overrides in the vast majority of cases any appeal to self-interest on the job and (2) easily threatened by self-interest (given human nature) in a way which warrants especially strong safeguards to ensure that it is complied with by employees.

As a final point, we should note that even the appearance of seeking a personal profit at the expense of one's employer is considered unethical since appearances of wrong-doing can harm a corporation as much as any actual biasing that might result from such practices. For example, using inside information to gain a personal advantage for oneself or one's family may not directly hurt a company—indeed, it directly harms only those who are thereby denied a fair opportunity to compete for the advantage. But if such activities become generally known, the company's public image can be hurt—a state of affairs no employer wants to see.

Study Questions

1 "Facts: Engineer Doe is employed on a full-time basis by a radio broadcast equipment manufacturer as a sales representative. In addition, Doe performs consulting engineering services to organizations in the radio broadcast field, including analysis of their technical problems and, when required, recommendation of certain radio broadcast equipment as may be needed. Doe's engineering reports to his clients are prepared in form for filing with the appropriate governmental body having jurisdiction over radio broadcast facilities. In some cases Doe's engineering reports recommend the use of broadcast equipment manufactured by his employer."

"Question: May Doe ethically provide consulting services as described?" (*NSPE Opinions of the Board of Ethical Review*, Case No. 75.10)

2 "Henry is in a position to influence the selection of suppliers for the large volume of equipment that his firm purchases each year.

"At Christmas time, he usually receives small tokens from several salesmen, ranging from inexpensive ballpoint pens to a bottle of liquor. This year, however, one salesman sends an expensive briefcase stamped with Henry's initials" (Kohn and Hughson, 104).

Should Henry accept the gift? Should he take any further course of action?

3 Read the case study at the beginning of Chap. 6. Was Brown and Root Corporation caught in a conflict of interest by being both the original designer of the road and the subsequent overseer of construction? If so, was the conflict permissible?

CONFIDENTIALITY

Many instances of failing to protect confidential information qualify as conflicts of interest. Nevertheless, the necessity of protecting such information is a distinct obligation of engineers and important in its own right. Indeed, keeping confidences is one of the most central and widely acknowledged duties of any professional. Defense attorneys must keep information clients tell them confidential, doctors and counselors must keep information on their patients confidential, and so too employed engineers must keep privileged information about their companies and their clients confidential.

Definition

What sort of information are engineers obligated to keep confidential? Two questions are really involved here. First, what is meant by the term "confidential information?" Second, how specifically can we identify what data should be kept confidential?

The first question is easier to answer. Confidential information is information which is to be kept secret. "Kept secret" is a relational expression. It always makes sense to ask, "Secret with respect to whom?" In the case of some government organizations, such as the FBI and CIA, highly elaborate systems for classifying information have been developed that identify which individuals and groups may have access to what information. Within other governmental agencies and private companies, engineers and other employees are usually expected to withhold information labeled "confidential" from unauthorized people both inside and outside the organization.

The second question, which concerns the criteria for identifying what information should be treated as confidential, is somewhat more difficult to answer. One criterion is suggested in the code of ethics of the former Engineers' Council for Professional Development: "Engineers shall treat information coming to them in the course of their assignments as confidential" (Sec. 4-i). But this is too broad. Some of the information acquired on assignments is routine and widely known. For example, it may be knowledge about new company facilities or plans which is readily available to anyone. Or while working on a project an engineer may become familiar with technical processes known generally throughout the industry.

A different criterion would identify any information which if it became known would cause harm to the corporation or client. Yet there are always questions about just what information would produce that result. Being able to give a precise answer would require the talents of a fortuneteller.

Most businesses tacitly adopt yet another criterion: Confidential information is any information which the employer would like to have kept secret in order to compete effectively against business rivals. Often this is understood to be any data concerning the company's business or technical processes which are not already public knowledge. While this criterion is somewhat vague, it clearly points to the employer as the main source for deciding what information is to be treated as confidential. It is the criterion we will adopt.

Related Terms

Several related terms need to be distinguished. *Privileged information* is an expression often used as a synonym for "confidential information." Literally it means "available only on the basis of special privilege," such as the privilege accorded an employee working on a special assignment. It covers information which has not yet become public or widely known within an organization.

Proprietary information is information which a company owns or is the proprietor of. This term is used primarily in a legal sense just as "property" and "ownership" are ideas carefully defined by law. Normally it refers to new knowledge generated within the organization which can be legally protected from use by others.

A rough synonym for proprietary information is *trade secrets*. A trade secret can be virtually any type of information which has not become public and which an employer has taken steps to keep secret. It may be data about designs and technical processes, organization of plant facilities, quality control procedures, customer lists, business plans, and so on (Popper, 101). Trade secrets are given limited legal protection against employee abuse. They are protected by common law—law generated by previous court rulings—rather than by statutes passed legislatively. An employer can sue employees for divulging trade secrets, or even for planning to do so. To win such a case, the employer must be able to prove the information had been or is being actively protected (for example, by showing it was or is available only to special employees for specific purposes, that contracts require subcontractors to keep the data secret, and so forth).

Patents differ from trade secrets. Patents legally protect specific products from being manufactured and sold by competitors without the express permission of the patent holder. Trade secrets have no such protection. A corporation may learn about a competitor's trade secrets through legal means such as "reverse engineering," where an unknown design or process can be traced out by analyzing the final product. But patents do have the drawback of being public and thus allowing competitors an easy means of working around them by finding alternative designs. They also can be held for only 17 years, whereas trade secrets, so long as they can be kept secret, have no time restrictions.

Patents are protected by statute laws passed in order to provide incentives for creativity (Vaughn, 34). In effect they give the patent holder the reward of a legally protected monopoly. By contrast, the legal protection accorded trade secrets is limited to upholding relationships of confidentiality and trust.

Moral Basis of the Confidentiality Obligation

Upon what moral basis does the confidentiality obligation rest, with its wide scope and obvious importance? Specifically, why are employers allowed to determine what information is to be treated as confidential? And what are the moral limits or restrictions on the confidentiality obligations of employees?

The major ethical theories can be applied to answer these questions. Advocates of every theory would probably agree that employers have some moral and institutional rights to decide what information relating to their organizations can be released publicly. They acquire these rights as part of their charge to protect the interests of their organizations—whether those interests be a company's competitive edge (in the case of profit-making institutions) or the safety and well-being of clients or consumers. In addition, in the case of information like trade secrets developed by the company, there can be a right of ownership of the intellectual property (Schwarze). But different ethical theories will justify the rights differently and will also differ in the limits they place on them.

Briefly, a *rights-based* theorist will appeal to more basic considerations: for example, the right of stockholders to have management pursue a course consistent with their own best interests and the general rights of property ownership. This right, in turn, might be grounded in the fundamental moral right of the stockholders to pursue their legitimate interests within a socially accepted free-enterprise system. However, the right of employers to establish what information should be treated as confidential will be limited by other legitimate moral rights: Minimally, no employer has a right to prevent engineers from whistleblowing in cases where public knowledge of information would save human lives and thereby protect the rights of people to live.

Duty-based theorists will emphasize the basic duties of both employers and employees to maintain the trust placed in them at the time they committed themselves to an employment agreement, a commitment that is understood to extend beyond the time of actual employment. They may also appeal to general duties not to abuse the property of others. Such duties, though, can be overridden by others, such as the duty to protect innocent lives that might occasionally require whistleblowing.

Utilitarians will view the authority of employers to determine the rules governing confidentiality as justified to the extent that it produces the most good for the greatest number of people. What this extent might be will depend on the particular theory of goodness subscribed to and the means required in any given situation to produce the most good. *Act-utilitarians* will focus on each instance where an employer decides on what is to count as confidential information. Is that act the most beneficial for everyone affected by it? *Rule-utilitarians,* by contrast, will emphasize the general benefits that result from having rules to protect confidential information. For example, investors' profits benefit from guarding trade secrets from competitors, and all society benefits from a system of limited protection of secrets to the extent that it stimulates creation of alternative products. The limits of the confidentiality obligation for utilitarians will depend on when acts or rules of keeping information confidential do not produce the best consequences.

Confidentiality and Changing Jobs

The obligation to protect confidential information does not cease when employees change jobs. If it did, it would be impossible to protect such information. Former employees would quickly divulge it to their new employers, or perhaps for a price sell it to competitors of their former employers. Thus the relationship of trust between employer and employee in regard to confidentiality continues beyond the formal period of employment. Unless the employer gives consent, former employees are barred indefinitely from revealing trade secrets. This provides a clear illustration of how the professional integrity of engineers involves much more than mere loyalty to one's present employer.

Yet thorny problems arise in this area. Many engineers value professional advancement more than long-term ties with any one company and so change jobs frequently. Engineers in research and development are especially likely to have high rates of job turnover. They are also the people most likely to be exposed to important new trade secrets. Moreover, when they transfer into new companies they frequently do the same kind of work as before—precisely the type of situation where trade secrets of their old companies may have relevance.

Donald Wohlgemuth and B.F. Goodrich   Consider, for example, the case of Donald Wohlgemuth, a chemical engineer who at one time was manager of B.F. Goodrich's space suit division (Baram, 208). Technology for space suits was undergoing rapid development, with several companies competing for government contracts. Dissatisfied with his salary and the research facilities at B.F. Goodrich, Wohlgemuth negotiated a new job with International Latex Corporation as manager of engineering for industrial products. International Latex had just received a large government subcontract for developing the Apollo astronauts' space suits, and that was one of the programs Wohlgemuth would manage.

The confidentiality obligation required that Wohlgemuth not reveal any trade secrets of Goodrich to his new employer. But this is easier said than done. Of course it is possible for employees in his situation to refrain from explicitly stating processes, formulas, and material specifications. Yet in exercising their general skills and knowledge, it is virtually inevitable that some unintended "leaks" will occur. An engineer's knowledge-base generates an intuitive sense of what designs will or will not work, and trade secrets form part of this knowledge-base. To fully protect the secrets of an old employer on a new job would thus virtually require destroying part of an engineer's brain—a solution no one would recommend on moral grounds!

Is it perhaps unethical, then, for employees to make job changes in cases where unintentional revelations of confidential information are a possibility? Some companies have contended that it is. Goodrich, for example, charged Wohlgemuth with being unethical in taking the job with International Latex. Goodrich also went to court seeking a restraining order to prevent him from working for International Latex or any other company which developed space suits. The Ohio Court of Appeals refused to issue such an order, although it did issue an injunction prohibiting Wohlgemuth from revealing any Goodrich trade secrets. Their reasoning was that while Goodrich had a right to have trade secrets kept confidential, it had to be balanced against Wohlgemuth's personal right to seek career advancement. And this would seem to be the correct moral verdict as well.

Management Policies

What might be done to recognize the legitimate personal interests and rights of engineers and other employees while also recognizing the rights of employers in this area? And how can obligations to maintain confidences of former employers be properly balanced against obligations to faithfully serve the interests of new employers? There are no simple answers to these questions. Difficult dilemmas will always arise which call for sensitive and creative moral judgment. But while the law has not found it wise to pass strict legislation in this complicated area, some general management policies are being explored (Baram, 212–215).

One approach is to use employment contracts that place special restrictions on future employment. Traditionally those restrictions have centered on geographical location of future employers, duration of time after leaving the present employer before one can engage in certain kinds of work, and the type of work it is permissible to do for future employers. Thus Goodrich might have required as a condition of employment that Wohlgemuth sign an agreement that if he sought work elsewhere he would not work on space suit projects for a competitor in the United States for 5 years after leaving Goodrich.

Yet such contracts are hardly agreements between equals, and they threaten the right of individuals to pursue their careers freely. For this reason the courts have tended not to recognize such contracts as binding, although they do uphold contractual agreements forbidding disclosure of trade secrets.

A different type of employment contract is perhaps not so threatening to employee rights in that it offers positive benefits in exchange for the restrictions it places on future employment. Consider a company which normally does not have a portable pension plan. It might offer such a plan to an engineer in exchange for an agreement not to work for a competitor on certain kinds of projects for a certain number of years after leaving the company. Or another clause might offer an employee a special post-employment annual consulting fee for several years on the condition that he or she does not work for a direct competitor during that period.

Other tactics aside from employment contract provisions have been attempted by various companies. One is to place tighter controls on the internal flow of information by restricting access to trade secrets except where absolutely essential. The drawback to this approach is that it may create an atmosphere of distrust in the workplace. It might also stifle creativity by lessening the knowledge-base of engineers involved in research and development.

There have been gentlemen's agreements among competing corporations not to hire one another's more important employees. But the problem here is that when such practices become widespread in a given industry, some of the best engineers may be turned away to other fields offering greater job options.

One potential solution is for employers to help generate a sense of professional responsibility among their staff that reaches beyond merely obeying the directives of current employers. Engineers then can develop a real sensitivity to the moral conflicts they may be exposed to in certain job changes. They can arrive at a greater appreciation of why trade secrets are important within a competitive system and learn to take the steps necessary to protect them. In this way professional concerns and employee loyalty can become intertwined and reinforce each other.

Study Questions

1 Consider the following example:

Who Owns Your Knowledge? Ken is a process engineer for Stardust Chemical Corp., and he has signed a secrecy agreement with the firm that prohibits his divulging information that the company considers proprietary.

Stardust has developed an adaptation of a standard piece of equipment that makes it highly efficient for cooling a viscous plastics slurry. (Stardust decides not to patent the idea but to keep it as a trade secret.)

Eventually, Ken leaves Stardust and goes to work for a candy-processing company that is not in any way in competition. He soon realizes that a modification similar to Stardust's trade secret could be applied to a different machine used for cooling fudge, and at once has the change made (Kohn and Hughson, 102).

Has Ken acted unethically? Defend your view.

2 Answer the following questions asked by Philip L. Alger, N.A. Christensen, and Sterling P. Olmsted in their book *Ethical Problems in Engineering*:

If an engineer has been unjustly discharged, must he keep confidential in later employment the trade secrets of his original employers? In general, is it wise to follow the doctrine of an "eye for an eye and a tooth for a tooth"? (Alger, Christensen, and Olmsted, 111).

3 Alger, Christensen, and Olmsted also give the following example:

Client A solicits competitive quotations of the design and construction of a chemical plant facility. All the bidders are required to furnish as a part of their proposals the processing scheme planned to produce the specified final products. The process generally is one which has been in common use for several years. All of the quotations are generally similar in most respects from the standpoint of technology.

Contractor X submits the highest-price quotation. He includes in his proposals, however, a unique approach to a portion of the processing scheme. Yields are indicated to be better than current practice, and quality improvement is apparent. A quick laboratory check indicates that the innovation is practicable.

Client A then calls on Contractor Z, the low bidder, and asks him to evaluate and bid on the alternate scheme conceived by Contractor X. Contractor Z is not told the source of alternate design. Client A makes no representation in his quotation request that replies will be held in confidence.

Is Client A justified in his procedure? (Alger, Christensen, and Olmsted 177).

4 American Potash and Chemical Corporation advertised for a chemical engineer having industrial experience with titanium-oxide. It succeeded in hiring an engineer who had formerly supervised E.I. du Pont de Nemours and Company's production of titanium-oxide. Du Pont went to court and succeeded in obtaining an injunction prohibiting the engineer from working on American Potash's titanium-oxide projects. The reason given for the injunction was that it would be inevitable that the engineer would disclose some of du Pont's trade secrets (Carter, 54). Defend your view as to whether the court injunction was morally warranted or not.

5 The importance of employee confidentiality was recently underscored by cases of industrial espionage in "Silicon Valley" (Santa Clara County, California). One of these cases even involved alleged purchases of confidential information by employees of Japanese competitors. Research some of these cases and report on the moral issues involved. You may want to refer to articles by Hiltzik and Samuelson listed in the Bibliography.

UNIONISM

Is it possible for an engineer to be a professional, dedicated to the highest ethical standards of professional conduct, while simultaneously being a member and supporter of a union? The question, we feel, is too complex to warrant a simple answer. Before answering it we would need to know what kind of union and union activities are at issue. Lacking this information, the answer would seem to be: sometimes yes and sometimes no.

Yet many observers have argued that the ethical aspects of professionalism in engineering are inherently inconsistent with unionism—that is, with union ideology and practice. In *The Engineer and His Profession*, for example, John Kemper writes:

There is little doubt that unionism and professionalism are incompatible. Professionalism holds that the interests of society and of the client (or employer) are paramount. Unions are collective bargaining agents that sometimes place the economic interests of the members ahead of those of the client or employer (Kemper, 253).

A number of professional societies have also held that loyalty to employers and the public is incompatible with any form of collective bargaining. The National Society of Professional Engineers (NSPE) has fervently led the opposition to union organizing of engineers (Seidman, 224). Its position is reflected in the NSPE Code of Ethics: "Engineers shall not actively participate in strikes, picket lines, or other collective coercive action" (Section III, 1e). Before discussing two arguments for this view, let us take note of a few historical facts about unions and engineering.

Historical Note

The beginnings of engineering unionism in the United States occurred during World War I (Seidman, 229). Marine architects and drafters were dissatisfied with their salaries at a time of both rising living costs and rising wages of blue-collar workers. Various groups were organized which later unified to become the American Federation of Technical Engineers, an affiliate of the former American Federation of Labor (AFL).

Most contemporary engineering unions, however, had their origin during the 1940s. These groups usually remained independent of the large national unions like AFL and CIO (Congress of Industrial Organizations). World War II and its aftermath brought widespread job insecurity, unhappiness with salaries, and lessened professional recognition (Walton, 18–45). Yet engineering unions were never able to organize most engineers. In fact, at their peak during the late 1950s, engineering unions had only 10 percent of the total number of engineers as members.

Beginning around 1960, what unionism there was in engineering declined, although a number of relatively small unions are active today. Several explanations have been offered for this decline (Seidman, 220). Certainly one major factor was the generally steady rise in economic well-being of engineers. During the past two decades engineering salaries have risen favorably in comparison with salaries of graduates of other 4-year programs (although they have not necessarily risen in terms of real income), and companies have sought to improve other job benefits. With many new technologies developing, moreover, engineers have been in high demand. Thus it is not surprising that many engineers surveyed have said they are opposed to unions because they think them unnecessary (Riegel, 43).

These observations do not necessarily apply to all industries. In the aerospace industry, for instance, a history of high job turnovers has created a highly mobile group of engineers with lessened job security. When engineers at two major aerospace firms were polled in a study by Archie

Kleingartner, 30 percent of those eligible to join were found to be members of unions. While this is by no means a majority, a surprising number of engineers working for the two firms disagreed with the statement that "it is impossible for an engineer to belong to a union and at the same time to maintain the standards of his profession." The percentages disagreeing ranged from 68 percent among low-level professionals to 91 percent among high-level professionals (Kleingartner, 230). As a result of his study, Kleingartner concluded,

> . . . the majority of engineers interviewed . . . do not view unionism as threatening their professionalism, and very likely also they do not see it coming between them and management in any fundamental way. They attribute substantially less importance to the potentially disrupting effects of unions than does management. The engineers view unions as limited institutions performing certain limited functions (Kleingartner, 235).

Two further reasons might be mentioned for why there is little unionism in engineering. First, in many industries engineers increasingly have come to identify with management. A degree in engineering together with several years of work experience opens the door to becoming a manager. Knowing this would not incline engineers to support union groups which often stand in adversary relationships with management.

Second, the NSPE and other engineering societies have for a long time been strongly opposed to unionism on the grounds that it is unprofessional. They have actively campaigned in opposition to attempts to unionize engineers.

Let us now turn to two arguments which support this stand taken by the NSPE: The first we will call the "faithful agent argument" and the second the "public service argument."

The Faithful Agent Argument

In the current NSPE Code the ban on the use of "collective coercive action" appears as one of the principles of obligation concerning professional integrity (Section III 1-e). Yet in earlier versions it was placed prominently in the first section which dealt with loyalty to employers:

> Section 1—The Engineer will be guided in all his professional relations by the highest standards of integrity, and will act in professional matters for each client or employer as a faithful agent or trustee. . . .
> f. He will not actively participate in strikes, picket lines, or other collective coercive action (1979 NSPE Code).

The implication is that being the faithful trustee of one's employer is incompatible with actively supporting collective action aimed against that employer.

In a number of NSPE publications this position has been explicitly endorsed. In 1976, for example, NSPE's Board of Ethical Review reiterated it in discussing a hypothetical example (Case No. 74-3). The case concerned the unionized employees in a state highway department. The employees, most of whom were not engineers, voted to strike when their demands for a pay increase of 60 percent and other benefits were denied. The Board of Ethical Review insisted that it was unethical for the engineers to participate actively, even though not to do so might mean facing union penalties. Passive participation, such as not crossing picket lines, was ruled permissible if it was necessary to avoid physical danger or abuse. The argument given was concise: "the engineers have a higher standard than self-interest; they have the necessary ethical duty to act for their employer as a faithful agent or trustee."

Obviously the Board saw active support of a strike or other collective action used against an employer as a violation of professional *ethics*, which it identified with the duty engineers have to serve as their employer's "faithful agents or trustees." Many people involved in engineering would agree with such a view, and certainly a case can be made for it. The conduct under discussion involves several features, any one of which might seem inconsistent with loyalty to employers: (1) It goes against the desires or interests of the employer, (2) it uses coercion or force against the employer, and (3) it involves *collective* and organized opposition. Certainly we can all think of behavior along these lines which is unprofessional and disloyal. The difficulty, however, is that not every instance of such conduct is unethical, as the following two examples show.

Consider three supervisory engineers who have good reason to believe they are being underpaid. After individually reasoning with their bosses to no avail they threaten—in a polite way—to seek employment elsewhere. In doing so, they act against the desires and interests of their employer, and they use a type of collective coercion. But they have not acted unethically or violated their duty to their employers. The point, which should by now be familiar, is that the duty to an employer has limits. Loyalty and faithfulness do not always require sacrificing one's own self-interest to an employer's business interests.

Or consider this second case: Management at a mining and refinery operation have consistently kept wages below industry-wide levels. They have also sacrificed worker safety in order to save costs by not installing special structural reinforcements in the mines, and they have made no effort to control excessive pollution of the work environment. As a result the operation has reaped larger than average profits. Management has been approached by both individuals and representatives of employee groups about raising wages and taking the steps necessary to ensure worker safety, but to no avail. A nonviolent strike is called and the metallurgical engineers support it for reasons of worker safety and public health. Here collective action aimed at coercing an employer is being used—and specifically a strike. But is it unethical or unprofessional?

It will be objected that these are special cases, and of course they are. They were designed expressly to show that it is not always obvious that strikes or other collective, forceful, action on the part of employees is unprofessional, excessively self-interested, or disloyal to employers. One must look at specific unions, specific strikes, specific situations. Even in the case discussed by the Board of Ethical Review we would want to know whether the union demands were reasonable. Were the workers so seriously underpaid that a 60 percent raise was not as fantastic as it sounds? And did the other benefits demanded relate to worker safety, compensation for injury, or possibly even public highway safety?

The examples suggest two weak generalizations. First, employee duty to employers does not entail unlimited sacrifice of economic self-interest. "Faithful agency" primarily concerns carrying out one's assigned tasks; it does not mean that one should never negotiate salary and other economic benefits from a position of strength.

Second, as the NSPE Code itself states, the duty to employers is limited by the more paramount duty to protect public health, safety, and welfare. Moreover, duty to employers is also limited by considerations such as worker safety and the right to refuse to obey illegal or unethical directives. Collective action of a coercive nature might sometimes be the only effective way to pursue these concerns of overriding importance. Professional societies have themselves engaged in a type of collective, coercive, action when they print editorials in official journals exposing companies for abuses they have committed against engineers.

NSPE recommends the use of a sounding-board, composed of a mix of employees and managerial engineers, to settle disputes with employees through reasonable dialogue. Certainly where feasible this is preferable to the use of collective force. Yet it would take considerable optimism to think that this procedure will always provide adequate support for salaried engineers.

The Public Service Argument

A second general argument against unions begins by emphasizing that the paramount duty of engineers is to serve the public. It then notes that by definition unions seek to promote the special interests of their members, not the interests of the general public. It is inevitable, so the argument continues, that clashes will occur, posing a threat to the meeting of professional commitments to the public. Strikes, which are the ultimate source of power for unions, may wreak havoc with the public good. Witness what has happened in recent strikes by police officers, fire fighters, teachers, and nurses. Then imagine what would happen to the economy if all computer engineers and technicians were to go on strike!

There is force in this argument. Yet once again it points out only the dangers of unions, even using the worst possible scenario of what might happen, and assumes that engineering unions must act irresponsibly. Of course many unions have acted in that way, but not all.

It is at least possible that a collective bargaining group for engineers, whether called a union, a guild, or an association, led by professional engineers, could devote itself to promoting the interests of engineers only within the limits set by professional concern for the public good. It could also devote itself to giving positive support to ethical conduct by engineers —which, after all, is part of the self-interest of morally concerned engineers. As we shall see more fully in Chap. 6, engineers who have sought to protect the public have not always fared well at the hands of management. The collective power of a guild or union might prevent the vindictive firing of responsible whistleblowers (Shapley, 620). It might also secure certain economic benefits, such as portable pensions, which would allow engineers a greater measure of freedom to act in the face of possible dismissal for reasons of whistleblowing or refusing to act unethically.

Conclusion

What we have said is neither a general endorsement of unionism nor a blanket condemnation of it. Our concern was the limited one of questioning two of the main arguments used to show that there is an inherent inconsistency between professionalism and unionism. Whether collective bargaining and its tactics are unethical or not depends on the details of any given situation involving their use.

We would agree that unions often enough have abused their power and irresponsibly disregarded the public good, so that the formation of any new union carries with it new risks to professionalism. But to conclude, therefore, that the formation of engineering unions is *always* unprofessional is like arguing that because a new technology involves risks it should never be developed.

The moral assessment of unions is complex, and a considerable number of morally relevant facts must be considered before a judgment can be made about any specific case or before a generalization can be formed. Disputes over unionism itself, however, typically involve disagreements over claims like those given in the following two lists (Burton, 129; Kemper, 249–257).

Union Critics

1 Unions are a main source of the inflation which is devastating the economy. Unions harm the economy by placing distorting influences on efficient uses of labor.

2 Unions encourage adversary, rather than cooperative decision making. They also remove person-to-person negotiations between employers and employees and make the individual worker a pawn of the collective bargaining group.

3 Unions promote mediocrity and discourage initiative by emphasizing job security and by making job promotion and retention rest on seniority. Management is prevented from rewarding individuals by having to negotiate salaries according to job description and length of company service rather than according to personal achievement. A further side effect is the pigeon-holing of employees in narrow job classifications to which the salary scales are attached.

4 Unions encourage unrest and strained relations between workers and management.

Union Supporters

1 Unions have been the primary factor in creating healthy salaries and the high standard of living enjoyed by today's workers. Even nonunionized workers have benefited since their employers must pay salaries comparable to those unions win for their workers.

2 Unions give employees a greater sense of participation in company decision making. For example, the European practice of codetermination in which union representatives serve on boards of directors has often created labor peace.

3 Unions are a healthy balance to the power of employers to fire at will. They give workers greater job security and protection against arbitrary treatment. Employees with union backing are more able to resist orders to perform unethical acts.

4 Unions yield stability by providing an effective grievance procedure for employee complaints. They are also a counterforce to radical political movements which exploit worker dissatisfaction and alienation.

Study Questions

1 Present and defend your view as to whether collective action aimed against employers always involves unfaithfulness and disloyalty on the part of the employees. In doing so distinguish between the two senses of "loyalty" given earlier in this chapter. Consider issues like salary, harmful labor practices, and the public good. Also consider the use of collective action by different groups, such as (a) unions, (b) professional societies, (c) nonunion employee groups, and (d) manufacturer's associations and trade organizations.

2 Answer the questions asked by Philip M. Kohn and Roy V. Hughson in regard to the following case. Give reasons.

Reginald's company pays its engineers overtime plus a bonus to work during a strike. The plant is being struck over "unsafe" working conditions, a claim that the company disputes. Reginald, considered by the company to be "management," believes conditions *may* be unsafe, even though no government regulations apply. Should Reginald:

    1. Refuse to work, because he thinks the union's allegations may have merit?
    2. Refuse to work, because he believes that strike-breaking is unethical?
    3. Work, because he feels this is an obligation of all members of management?
    4. Work, because it is a great way to catch up on some of his bills, or earn the down payment on a car, etc.?
    5. Work, because he believes he may be fired if he doesn't?
    6. Other? (Please specify) (Kohn and Hughson, 102 and 105)

SUMMARY

Professions are occupations requiring sophisticated skills, extensive formal education, group commitment to some public good, and a significant degree of self-regulation. *Persuasive definitions* are frequently given for terms like "professionalism," "professional," and "profession." That is, special criteria (cognitive meaning) are applied to the terms with their generally positive connotations (positive emotive meaning). For example, some people think of the "professionalism of salaried engineers" as being centered in loyal service to employers, while others have seen it as freedom from control by employers; both, however, are persuasive definitions in that they link the emotional connotations of a term to special (and in this case controversial) cognitive criteria for applying the term.

In our view, the duty of engineers to the public is paramount in the sense that it deserves special emphasis given the contemporary obstacles to meeting that duty. Yet it is too much to say that obligations to the public always and everywhere should override obligations to employers. Both obligations are important. When they come into conflict it is necessary to examine the specific situation before deciding which ought to take precedence.

The relationship between loyalty to employers and other professional obligations is complex. Loyalty to employers can mean (1) meeting one's moral obligations to employers—in which case loyalty is automatically good; (2) being zealously supportive of the employers' interests—in which case there are limits to how far loyalty is good.

Institutional authority involves the right to exercise power based on the resources of the institution. It allows people to meet their institutional duties, and the prerogatives it entails are specified by rules designed to further the institution's good; it is not the same as *expert authority* (special knowledge or expertise). Institutional authority is morally justified only where the goals of the institution are morally permissible or desirable and when the way in which it is exercised does not violate other moral duties.

Authority relationships between employers and employees are normally necessary for avoiding the negative effects of unlimited individual discretion. And the employment contract constitutes a promise on the part of employees to recognize legitimate institutional authority. The obligation to obey authoritative directives, however, should not be construed as an obligation to suspend one's critical faculties and blindly follow those directives regardless of their moral content.

Employee conflicts of interest occur when employees have side interests which if pursued could prevent them from meeting their obligation to serve the interests of their employers. Such side interests are generally understood to threaten employer interests in one main way: They have the potential to bias the employee's independent judgment. Examples of conflicts of interest include moonlighting for a competitor, misusing inside confidential information for personal gain, and accepting substantial gifts from clients or suppliers. Some conflicts of interest are permissible, however, subject to the employer's approval.

Confidential information is information which an employer or client judges should be kept secret to serve the company's or client's interests. *Proprietary information* and *trade secrets* are information which is protected by the courts. The confidentiality obligation can be justified within rights-based theories (for example, by reference to the rights of stockholders or the rights to intellectual property of corporations), within duty-based theories (by reference to the mutual promises of the employment contract), and within utilitarian theories (by reference to the benefits derived by companies and the public). Moral dilemmas can arise for engineers when they move to new jobs since they may possess privileged information from their old jobs which they carry with them. The confidentiality obligation extends beyond the old job, however, and places reasonable restraints on engineers in regard to how and when they may work for new employers. The confidentiality obligation is limited by the public's right to be warned of potential hazards.

Unionism and professionalism seem inherently incompatible when the duty of employees to employers is seen as paramount and unlimited. But when that duty is viewed as limited by both a legitimate degree of self-interest on the part of employees and the wider good of the public, the incompatibility becomes less clear. Rather, individual unions and union tactics must be assessed in terms of their positive and negative effects in the situations in which they arise.

CHAPTER 6

RIGHTS OF ENGINEERS

In 1966 Charles Pettis was sent by Brown and Root Overseas, Inc., to serve as resident engineer in Peru. Brown and Root had been hired by Peru's government to protect its interests on a project being undertaken by another firm, Morrison-Knudsen. The project was breathtaking: construction of a 146-mile highway across the Andes Mountains. The highway would open major new trade routes between Peru's coast and its isolated inner cities on the other side of the Andes. At age forty-four, with years of experience as a geological engineer behind him, Pettis was given the key assignment of ensuring that contract agreements between the Peruvian government and Morrison-Knudsen were met. His signature on the payroll certified that the interests of the Peruvian government were being served.

Almost immediately Pettis experienced doubts about the project. The design for the highway, which had originally been done by Brown and Root and was therefore a source of potential conflict of interest, called for cutting deep channels, some of them 300 feet deep, through the mountains with cliffs rising sharply on both sides of the road. Unfortunately, the Andes Mountains are known for their instability, and not enough geological borings had been taken to identify potential slide areas. Pettis's worries about this problem were confirmed when several slides and other construction incidents killed thirty-one workers.

Morrison-Knudsen instructed Pettis to add to the payroll in order to cover the substantial costs of slide removals. Pettis viewed this as padding and as not justified by anything in the contract. At first Brown and Root supported him. But by early 1968 Morrison-Knudsen had exerted sufficient pressure on Brown and Root management that they ordered Pettis to add the slide-removal costs to the payroll. He continued refusing to do so, insisting it would be a violation of the Peruvian government's interests he was charged with protecting. At that point Brown and Root relieved him of responsibility for payroll authorization.

Suspicions, however, had been aroused in the minds of Peru's transportation officials. They sought direct assurances from Pettis that the work was proceeding properly. Brown and Root placed enormous pressure on him to give those assurances, even promising him his pick of jobs if he cooperated. But Pettis refused to lie to his client. As a result Brown and Root fired him in December 1968.

In 1971 Pettis learned of Senator William Proxmire's investigations into the contract policies of multinational construction companies. He volunteered to testify before the General Accounting Office, and while doing so he blew the whistle on Brown and Root and Morrison-Knudsen. The General Accounting Office was able to confirm Pettis's charges of corporate misconduct (Peters and Branch, 183–186; Nader, Petkas, and Blackwell, 135–139).

Issues

Do engineers have a moral right to refuse to carry out what they consider to be unethical activity? How far are employers obligated to respect this right and to forgo the use of coercion and retribution in dealing with those employees who exercise it? Should engineers be recognized as having rights to speak out to clients, government regulators, and others concerning their employers' misconduct?

It may seem that endorsing such rights is incompatible with allowing employers full charge to direct a company. Isn't the position Pettis took inconsistent with recognizing management's rights? And what if management honestly disagrees with an engineer's safety judgments or interpretations of a contract?

Issues concerning the rights of engineers and other professionals working within organizations were usually given little attention. Only recently has the topic of the rights of employees been as seriously discussed as their duties and responsibilities. Indeed, the 1980s promises to be the decade when discussion of the rights of employed professionals reaches full maturity (Westin and Salisbury, 1980, xi).

PROFESSIONAL RIGHTS

Engineers have different types of moral rights which fall into the sometimes overlapping categories of human, employee, contractual, and professional rights. As human beings, engineers have fundamental rights to live and freely pursue their legitimate interests. For example, a human right we will discuss below in the section titled "Discrimination" is the right not to be unfairly discriminated against in employment on the basis of sex, race, or age. Another example, mentioned in connection with confidentiality in Chap. 5, is the human right to pursue one's career.

As employees, engineers have special rights, some of which will be explored later in the section titled "Employee Rights." Some of those include institutional rights which arise from specific agreements in the employment contract. For example, there is the right to receive one's salary and other company benefits in return for performing one's duties. However, other employee rights are not reducible to purely institutional rights. For example, the right to engage in the nonwork political activities of one's choosing, without reprisal or coercion from employers. Employers ought to respect this right, whether or not it is explicitly recognized in a contract or employment agreement.

Finally, engineers as professionals have special rights which arise from their professional role and the obligations it involves. Those include the right to form and express one's professional judgment freely (without intimidation), the right to refuse to carry out illegal and unethical activity, the right to talk publicly about one's work within bounds set by the confidentiality obligation, the right to engage in the activities of professional societies, the right to protect clients and the public from the dangers or harm that might arise from one's work, and the right to professional recognition (including fair remuneration) for one's services. All these, as we shall see, can be viewed as aspects of one fundamental professional right.

The Basic Right of Professional Conscience

There is one basic or generic professional right of engineers: the moral right to exercise responsible professional judgment in pursuing professional responsibilities. Pursuing those responsibilities involves exercising both technical judgment and reasoned moral convictions. For brevity, this basic right can be referred to as *the right of professional conscience.*

If the duties of engineers were so clear-cut that in regard to every situation it was obvious to every sane person what it was morally acceptable to do, there would be little point in speaking of "conscience" in specifying this basic right. Instead, we could simply say it is the right to do what everyone agrees it is obligatory for the professional engineer to do. But as we have seen throughout this book, engineering calls for as morally complex decisions as any other major profession does. It requires autonomous moral judgment in attempting to uncover the most morally reasonable courses of action, and the correct courses of action are not always obvious.

As with most moral rights, the basic professional right is an entitlement giving one the moral authority to act without interference from others. It is what we earlier called a "negative right," since it places an obligation on others *not* to interfere with its proper exercise.

Yet occasionally special resources may be required by the engineer seeking to exercise it in the course of meeting his or her professional obligations. For example, conducting an adequate safety inspection may require that special equipment be made available by employers. Or, more generally, in order to feel comfortable about making certain kinds of decisions on a project, the engineers involved may need an environment conducive to trust and support which management may be obligated to help create and sustain. In this way the basic right is also in some respects a "positive right," placing on others an obligation to do more than merely not interfere.

Institutional Recognition of Moral Rights

Having a moral right is one thing. Having it respected by others and given recognition within a corporation is quite another. When engineers appeal to the basic right of professional conscience they may be arguing for its institutional recognition by employers.

Consider in this connection the following comments made by two engineers at the 1975 Conference on Engineering Ethics:

> H. B. Koning: I think that one item that should be in the code of ethics is that engineers have the right at all times to exercise the dictates of their own consciences. For example, they need not apply their knowledge, skill and energy to scientific or technical business actions or plans which they feel will violate or lead to the violation of their personal or professional ethical standards (*Conference*, 99).
> N. Balabanian: Few engineers are self-employed. The vast majority work for others. What is desperately needed for engineer employees is to have a right of conscience. It isn't so much a matter of forcing engineers to conduct themselves ethically but to give them room—room for action—to carry out their own personal ethical convictions without threats of retribution (*Conference*, 101).

The first speaker is appealing to the moral right of professional conscience which engineers *do* have, even though it is not formally recognized in codes. He is arguing that this right should be stated formally and given official recognition. The second speaker seems to mean by a "right to conscience" an institutionally recognized right, one which engineers will have only after employers acknowledge and respect it. Both speakers are arguing for similar points, but using different language.

Specific Professional Rights

The right of professional conscience is the most basic, but also the most abstract, generic, professional right. It encompasses many other more particular rights. As with professional duties, specific professional rights can be stated in different ways involving different levels of generality.

For example, engineers have a general *obligation* to protect the safety and well-being of the public. Correspondingly, they have a general *right* to protect the safety and well-being of the public. As we will discuss in the next section, that obligation to the public might in special situations require whistleblowing. Thus engineers have a (limited) right to whistleblow. In turn, the whistleblowing right becomes more precisely specified by listing conditions under which whistleblowing is permissible. In general, as a particular professional obligation is more narrowly delineated, the corresponding professional right is also more precisely specified.

Realizing that professional rights can be stated with different degrees of abstraction helps us avoid two mistakes. First, just because some talk about professional rights is couched in abstract and general terms should not lead us to dismiss its significance. The same potential difficulty surrounds other rights. Consider the right to live. In the abstract, it sounds like it entails a right never to be killed. But it does not: for example, in situations where the only way to prevent a murder from occurring is to kill the murderer first. Such tacit limits on the right to live do not lead us to reject that right as nonsense. Similarly, sensitivity to the necessary limits on the rights of professional conscience within organizations should not lead us to dismiss lightly the importance of those rights.

A related second danger is that talk about rights may be used too loosely and not made specific with respect to given contexts. It will not do, for example, to object to every negative action by an employer as violating the rights of engineers. Even such vitally important rights as protection of public safety may in some situations be limited by the legitimate rights of employers—at least this possibility has to be explored. Neither the rights of engineers nor those of employers are unrestricted moral "passes," and there will always be difficult moral dilemmas involving conflicts between them. Such dilemmas can be resolved by developing cogent arguments for why one right should be limited in a specific context by another right.

Both the importance and the difficulty of applying professional rights in specific circumstances can be illustrated with the right of conscientious refusal and the right to professional recognition.

Right of Conscientious Refusal   The right of conscientious refusal is the right to refuse to engage in what one believes and has reason to believe is unethical behavior, and to refuse to do so solely because one views it as unethical. This is a kind of second-order right. It arises because other rights to pursue moral obligations within the authority-based relationships of employment sometimes come into conflict.

There are two situations to be considered: (1) where there is widely shared agreement in the profession as to whether or not an act is unethical and (2) where there is room for disagreement among reasonable people over whether an act is unethical.

It seems clear enough that engineers and other professionals have a moral right to refuse to participate in activities which are straightforwardly and uncontroversially unethical (e.g., forging documents, altering test results, lying, giving or taking bribes, or padding payrolls). And to coerce them into doing so by means of threats (e.g., to their jobs) plainly constitutes a violation of this right.

The troublesome cases concern situations where there is no shared agreement about whether or not a project or procedure is unethical. Possibly the Charles Pettis case involved different assessments of whether or not slide-removal charges could ethically be charged to Peru's government under the contract agreement. Do engineers have any rights to exercise their personal consciences in these more cloudy areas?

Let us approach this question with a rough analogy from medical ethics. There is no shared agreement over whether abortions are morally permissible or not. Yet, as is widely acknowledged, nurses who believe them to be immoral have a right to refuse to participate in abortion procedures. This is so even though nurses function under the institutional authority of doctors, clinics, and hospitals in ways analogous to how engineers work under the authority of management. Nevertheless, nurses' rights do not extend so far as to give them the right to work in an abortion clinic while refusing to play their assigned role in performing abortions.

Likewise, we believe engineers should be recognized as having a *limited* right to turn down assignments which violate their personal consciences in matters of great importance, such as threats to human life, even where there is room for moral disagreement among reasonable people about the situation in question. We emphasize the word "limited" because the right is contingent on the organization's ability to reassign them to alternative projects without serious economic hardship to itself.

For example, consider an engineer who requests not to work on a South African project because he views such work as supporting a racist regime. His corporation should be willing to try to find an alternative assignment for him, without any implication that he is being disloyal to the company. Yet if the bulk of the work for which he is needed concerns South African projects, he must be willing to seek employment elsewhere. The right of professional conscience does not extend to the right to be paid for not working.

Right to Recognition   Engineers have a right to professional recognition for their work and accomplishments. Part of this has to do with fair monetary remuneration, and part has to do with nonmonetary forms of recognition.

The right to reasonable remuneration is sufficiently clear to serve as a moral basis for arguments against corporations which make excessive profits while engineers are paid below pay scales of blue-collar workers. It can also serve as the basis for criticizing the unfairness of patent arrangements which fail to give more than nominal rewards to the creative engineers who make the discoveries leading to the patents. If a patent leads to millions of dollars of revenue for a company, it is unfair to give the discoverer a nominal bonus and a thank you letter.

But the right to professional recognition is not sufficiently precise to pinpoint just what a reasonable salary is or what a fair remuneration for patent discoveries is. Such detailed matters must be worked out cooperatively between employers and employees, for they depend upon both the resources of a company and the bargaining position of engineers.

It may seem, incidentally, that the right to fair remuneration merely concerns matters of self-interest, and as such does not properly fall under the basic right of professional conscience. Of course it does centrally involve self-interest. But there are also reasons why it is related to the basic right of conscience. For one thing, without a fair remuneration engineers cannot concentrate their energies where they properly belong—on carrying out the immediate duties of their jobs and on maintaining up-to-date skills through formal and informal continuing education. Their time becomes taken up by money worries, or even by moonlighting in order to maintain a decent standard of living. Or consider the seemingly "purely" economic issue of portable pensions. If a company's retirement plan is tied to on-going employment with that company, engineers will feel a considerable pressure not to leave their jobs. This pressure can deflect them from vigorously pursuing their obligations in situations where employers' directives are not in line with the legitimate needs or safety of clients and the public.

Nonmonetary forms of recognition are also important. Consider the following recent report by a 40-year-old chemical engineer:

I have had to write papers and sections of books which appeared under the authorship of my supervisor three levels up, on matters he can hardly understand, much less contribute to except by proof reading for grammatical errors. . . . The four key people whose work he became a world-recognized success by are disposed of as follows:
(1) Dead, heart attack, age 53, Ph.D. Chemical Engineering
(2) Dead, heart attack, age 42, M.S. Chemistry
(3) Dismissed from his job, age 49, Ph.D. Chemistry

(4) Mental breakdown, 2 months in psychiatric hospital, age 36, Ph.D. Chemical Engineering, currently seeking other employment (Bailyn, 73).

The point of this medical and obituary report is presumably to underscore how unhealthy it is to work hard at one's job without proper recognition. Unrecognized work is also demeaning. But just how far employers are morally required to go in providing fair recognition for their engineers is again a matter that must be regularly discussed and mutually agreed upon by management and engineers.

Moral Foundation of Professional Rights

Thus far we have said that engineers' professional rights, by definition, are those possessed by virtue of being engineers. More fully, they arise because of the special moral duties engineers acquire in the course of serving the public, clients, and employers. We have given several examples of those rights and illustrated the complexity of applying them. Next we must inquire into the moral basis or justification for asserting that such rights do indeed exist.

Professional rights and duties are not identical with the rights and duties of nonprofessionals, but neither are they unrelated to them. Professional rights and duties are justified in terms of more basic moral principles which also apply outside the professional job context. One's view of those more basic moral principles will depend, of course, on the particular ethical theory one endorses: rights-based, duty-based, or goal-based utilitarian.

There are two general ways to apply ethical theories in justifying the basic right of professional conscience. One is to proceed piecemeal by reiterating the justifications given for the specific professional duties. Whatever justification there is for the specific duties will also provide justification for allowing engineers the right to pursue those duties. Fulfilling duties, in turn, requires the exercise of moral reflection and conscience, rather than rote application of simplistic rules. Hence the justification of each duty ultimately yields a justification of the right of conscience with respect to that duty. But throughout this book we have illustrated how to justify various specific duties of engineers by means of more general ethical theories and there is no need to repeat that process here. Instead we shall pursue a second way to justify the right of professional conscience which involves grounding it more directly in the ethical theories. Here, as elsewhere, we invoke the ethical theories to serve as general models for organizing moral reflections and providing frameworks for approaching practical problems.

A Rights-Based Model Rights-based theories, it will be recalled, emphasize human moral rights as at least one ultimate ground of morality.

"Ultimate" means that human rights do not themselves need to be justified by referring to other, more fundamental moral principles. Thus a rights-based ethicist will seek to justify professional rights—in particular the basic right of professional conscience—by reference to human rights.

Let us follow A.I. Melden in viewing the most basic human right as the right to pursue one's legitimate interests. "Legitimate interests" will be those which do not violate others' rights. Hence the rights of any one individual must be understood within the context of a community of people, each of whom has rights which limit the extent of others' rights. Melden emphasizes that this community is a moral community, based upon ties of mutual understanding and concern (Melden, 1977, 140–145).

Although Melden does not himself apply his theory to professional rights, we would apply it as follows. "Legitimate interests" surely include moral concerns, especially concerns about meeting one's obligations. Thus the right to pursue legitimate interests implies a right to pursue moral obligations. This may be viewed as a human right of conscience directly derived from the most basic human right.

Now as engineers and other professionals take on special professional obligations, this general right of conscience acquires a further extension: It gives rise to a right of professional conscience which relates to specific professional obligations. In this way, the right of professional conscience is justified by reference to human rights as applied to the context of professional activity.

A Duty-Based Model   In duty-based ethics, rights are not the ultimate moral appeal. Instead they are mirror-image correlates of more basic duties. If I have a right to do something it is only because others have duties or obligations to allow me to do so. Within this context, the basic professional right is justified by reference to the duties others have to support or not to interfere with the work-related exercise of conscience by professionals. But who are these others, and what specifically do their duties entail? In regard to professionals, the "others" are their employers. And most importantly in regard to professional engineering, employers have a duty not to harm the public by placing handicaps in the way of the engineers they employ as those engineers seek to meet their obligations to the public. In addition, employers are directly obligated to professionals not to use coercion (i.e., not to threaten negative sanctions) which would encourage any compromise of personal moral integrity. To return to an earlier example, no hospital administrator has the right to pressure a Catholic nurse to participate in an abortion by threatening to fire her; to do so would show an utter disregard for her dignity as a moral agent (to use Kant's language). Similarly no employer has the right to threaten engineers with the loss of their jobs for refusing to work on projects they see as likely to lead to the death or injury of unsuspecting victims.

A Utilitarian Model Utilitarians will justify the right of professional conscience by reference to the basic goal of producing the most good for the greatest number of people. And no matter how "goodness" is defined, the public good is certain to be served by allowing professionals to meet their obligations to the public. For those obligations arise in the first place because of the role they play in promoting the public good.

Rule-utilitarians will seek to establish the best rule or policy in regard to employee rights for promoting the public good. Act-utilitarians will look at each situation to see whether and how far professionals should be allowed to exercise their consciences in pursuing their duties to the public.

Study Questions

1 Consider the following example by Philip M. Kohn and Roy V. Hughson:

> Jay's boss is an acknowledged expert in the field of catalysis. Jay is the leader of a group that has been charged with developing a new catalyst system, and the search has narrowed to two possibilities, Catalyst A and Catalyst B.
>
> The boss is certain that the best choice is A, but he directs that tests be run on both, "just for the record." Owing to inexperienced help, the tests take longer than expected, and the results show that B is the preferred material. The engineers question the validity of the tests, but because of the project's timetable, there is no time to repeat the series. So the boss directs Jay to work the math backwards and come up with phony data to substantiate the choice of Catalyst A, a choice that all the engineers in the group, including Jay, fully agree with (Kohn and Hughson, 103).
>
> What should Jay do, and does he have a moral right to not do as he is directed?

2 Comment on the following passage, making any suggestions about how engineers might be protected against such situations:

> Older engineers, in particular, find job security in competition with ethical instinct. With considerable sympathy, I recall the dilemma of an older PE, in the shadow of a comfortable retirement, who was confronted by a new general manager of the plant in which he was employed as a facilities engineer. In consideration of plans for a plant expansion, the general manager insisted that the PE reduce footings and structural steel specifications below standards of good practice. The PE was told to choose between his job and his seal on the plans. Did he really have a choice? (Howard, 48)

3 In 1971 Louis V. McIntire and Marion McIntire published a novel entitled *Scientists and Engineers: The Professionals Who Are Not*. The story was about the problems encountered by J. Marmaduke Glumm, a chemist working for the Logan Chemical Company. It portrayed the disillusionment of scientists and engineers pressured into becoming managers and thus forced to move away from their original areas of expertise. It also described the tactics management used to cheat employees out of bonuses, to show unjustified favoritism, to take unfair advantage

of employees in employment contracts, and to coerce professionals into going along with management's views on safety and health hazards. The novel recommended that engineers should form a national federation to seek laws protecting and favoring engineers working as employees.

The novel was a thinly disguised satire of the company Louis McIntire had worked for during the past 17 years: du Pont Company. When McIntire's employers learned about the novel in 1972 they fired him. In 1974 McIntire sued du Pont, but his claim that the First Amendment protected him from being fired was rejected by the courts.

Present and defend your view as to whether McIntire had a moral right not to be fired for writing the novel. Do you think the courts should have recognized such a right legally?

4 Leonardo da Vinci reported in his journal that he had discovered how to make what today we would call a submarine. He also noted that he refused to reveal the idea to anyone because of what he viewed as its likely misuse. He wrote:
. . . now by an appliance many are able to remain for some time under water. How and why I do not describe my method of remaining under water for as long a time as I can remain without food; and this I do not publish or divulge on account of the evil nature of men who would practice assassinations at the bottom of the seas, by breaking the ships in their lowest parts and sinking them together with the crews who are in them (da Vinci, 850).

Suppose that da Vinci discovered this idea while he was employed as a military engineer for Cesare Borgia or other military leaders, as he was at times in his career. Would he have had a moral right to refuse to reveal the idea to his employer? Would he be disloyal to the employer if he did refuse to reveal it? Defend your view by means of one of the ethical theories outlined earlier.

WHISTLEBLOWING

We have seen how obligations to the public and to clients may come into conflict with obligations to employers. And considering the importance of obligations to the public, especially the obligation to inform those members of the public affected by "social experimentation" through engineering, we have suggested that sometimes, though not always, obligations to the public override obligations to an employer. In seeking to meet those obligations to the public, engineers and others have sometimes engaged in what is known as "whistleblowing."

A variety of normative moral issues arise in connection with blowing the whistle on organizations: Is it ever morally permissible to do so? When? Is whistleblowing ever morally obligatory? Is it always an act of disloyalty to an organization, or could it sometimes be consistent with company loyalty—even an expression of it? Should it sometimes be viewed as an act of moral heroism which goes beyond the call of duty? What procedures ought to be followed in blowing the whistle? And to what extent do engineers have a right to whistleblow?

Before considering some of these questions, though, we need to define whistleblowing.

Definition of Whistleblowing

Whistleblowing is sometimes defined as making public accusations concerning misconduct by one's organization (Bok, 277; James, 99; Bowie, 142). This definition, however, is too narrow. On the one hand, an individual need not be a member of an organization in order to blow the whistle on it publicly. Journalists, politicians, and consumer groups may learn of corruption in organizations they do not work for and blow the whistle on them by publishing articles or informing regulatory agencies. Our main interest in this section, however, will be with whistleblowing by employees (both present and former employees), especially where disobedience of an employer's directives or company policies is involved.

On the other hand, not all whistleblowing involves going outside the organization. Recall the Ford engine-test case discussed at the beginning of Chap. 5. There the whistle was blown within the organization when the computer specialist wrote a memo to the company president informing him of misconduct in the engine and foundry division. Again, a case could be made for saying the BART engineers who blew the whistle to a member of BART's board of directors were operating within the organization since they intended only that the director should talk with BART's top management.

We shall not attempt to define all types of whistleblowing in all situations. Instead, we shall list four main features of most (though not all!) cases of whistleblowing by employees of organizations, whether the whistle is being blown on individuals or problems within the organizations:

1 Information is conveyed outside approved organizational channels or in situations where the person conveying it is under pressure from supervisors or others not to do so.

2 The information being revealed is new or not fully known to the person or group it is being sent to.

3 The information concerns what the whistleblower believes is a significant moral problem concerning the organization. Examples of significant problems are criminal behavior, unethical policies, injustices to workers within the organization, and threats to public safety.

4 The information is conveyed intentionally with the aim of drawing attention to the problem.

Using these four features as our definition, we will speak of *external whistleblowing* when the information is passed outside the organization.

Internal whistleblowing occurs when the information is conveyed to someone within the organization.

The definition also allows us to distinguish between open and anonymous whistleblowing. In *open whistleblowing* individuals openly reveal their identity as they convey the information. *Anonymous whistleblowing,* by contrast, involves concealing one's identity. But there are also overlapping cases, such as when individuals acknowledge their identities to a journalist but insist their names be withheld from anyone else.

Persuasive Definitions of Whistleblowing

Notice that the above definition leaves open the question of whether whistleblowing is justified or not. As we shall suggest in a moment, sometimes it is and sometimes it is not. By contrast, some writers have packed into their definitions of whistleblowing much of their own particular value perspectives concerning it. In doing so they have created persuasive or prescriptive definitions which sometimes blur the issues. Consider, for example, the following two proposals:

> "Whistleblowing"—the act of a man or woman who, believing that the public interest overrides the interest of the organization he [*sic*] serves, publicly "blows the whistle" if the organization is involved in corrupt, illegal, fraudulent, or harmful activity (Nader, Petkas, and Blackwell, vii).

> Some of the enemies of business now encourage an employee to be disloyal to the enterprise. They want to create suspicion and disharmony and pry into the proprietary interests of the business. However this is labelled—industrial espionage, whistleblowing or professional responsibility—it is another tactic for spreading disunity and creating conflict (Roche, 445).

The first definition was set forth by Ralph Nader at the beginning of a book which evaluates whistleblowing positively. Notice that the definition assumes that whistleblowing springs from an admirable motive: the belief that one is acting on behalf of the higher of two duties. It also assumes that whistleblowers hold accurate views about corporate wrongdoing. Thus in two ways it automatically implies a favorable general attitude toward whistleblowers.

The second passage was written by James M. Roche while he was chairman of the board of General Motors Corporation. It virtually identifies whistleblowing with motives like disloyalty and a malicious desire to harm the organization.

For the sake of clarity it is preferable to adopt a more value-neutral definition of whistleblowing, as we have done. Then the evaluative issues can be dealt with on their own merits.

Ernest Fitzgerald and the C-5A

One of the most publicized instances of open, external, whistleblowing occurred on November 13, 1968. On that day Ernest Fitzgerald was one of several witnesses called to testify before Senator William Proxmire's Subcommittee on Economy in Government concerning the C-5A, a giant cargo plane being built by Lockheed Aircraft Corporation for the Air Force. Fitzgerald, who had previously been an industrial engineer and management consultant, was then a deputy for management systems under the Assistant Secretary of the Air Force. During the preceding 2 years he had reported huge cost overruns in the C-5A project to his superiors, overruns which by 1968 had hit $2 billion. He had argued forcefully against similar overruns relating to other projects, so forcefully that he had become unpopular with his superiors. They pressured him not to discuss the extent of the C-5A overruns before Senator Proxmire's committee. Yet when Fitzgerald was directly asked to confirm Proxmire's own estimates of the overruns on that November 13, he told the truth.

Doing so turned his career into a costly nightmare for himself, his wife, and his three children (A. E. Fitzgerald; Peters, 200). He was immediately stripped of his current duties and assigned trivial projects, such as examining cost overruns on a bowling alley in Thailand. He was shunned by his colleagues. Within 12 days he was notified that his promised civil service tenure was a computer error. And within 4 months the bureaucracy was restructured so as to abolish his job. It took 4 years of extensive court battles before federal courts ruled that he had been wrongfully fired and ordered the Air Force to rehire him. And years of further litigation, involving fees of around $900,000, were required before, in 1981, he was reinstated in his former position.

Fitzgerald displayed remarkable courage at considerable self-sacrifice. Was he obligated to do what he did? The Code of Ethics for the United States Government Service says that employees should "put loyalty to the highest moral principles and to country above loyalty to persons, party, or government department" and that they should expose "corruption wherever discovered." A coverup of $2 billion of taxpayers' money in contract overruns would seem to qualify as corruption. Is the principle in the code a morally valid one? We believe it is. The alternative would be to endorse a kind of "organizational egoism" where only the good of one's particular in-group is emphasized.

If we feel any hesitation in saying Fitzgerald was obligated to whistleblow, it concerns whether it might be asking too much of someone in his position to do what he did. Perhaps it is beyond the call of duty to require such an incredible degree of personal sacrifice in performing one's job. In any case, his acts seem to us admirable to the point of heroism.

Not all whistleblowing, of course, is admirable, obligatory, or even permissible. Obligations to an organization are significant. As we have suggested, they are not automatically cancelled or outweighed by the obligation to the public in all situations. But Fitzgerald's case seems to us clear-cut because (1) he had made every effort to first seek a remedy to the abuses he uncovered by working within accepted organizational channels, (2) his views were well founded on hard evidence, and (3) the harm done to the Air Force by his disclosures was both a just treatment for its mismanagement of the C-5A project and far outweighed by the benefits that accrued to the public. In addition, (4) Fitzgerald was a public servant with especially strong obligations to the public which his organization, the Air Force, is committed to serve, and (5) to have withheld the information from Senator Proxmire would have involved lying and participating in a coverup. As is often true, failure to whistleblow in his case would have amounted to complicity in wrong-doing.

Carl Houston and Welding in Nuclear Plants

In 1970 Carl Houston was working for Stone and Webster, the contractor for a nuclear power facility being constructed in Surry, Virginia. Houston was assigned as a welding supervisor at the facility and immediately saw that improper welding procedures were being used. Wrong materials were being utilized and the welders had not been properly trained. The situation was especially dangerous since some of the defective welds were appearing on the water pipes carrying coolant to the reactor core. Rupture of the pipes could cause disaster if safety backups failed simultaneously.

Houston reported his observations to Stone and Webster's local manager, who disregarded them. When he threatened to write to Stone and Webster's headquarters, he was told he would be fired. He sought to alert the reactor suppliers about the danger, and shortly thereafter he was fired on the trumped-up grounds that he was not qualified for welding. Afterwards he wrote letters to the governor of Virginia and to the Atomic Energy Commission, which were never answered. Finally, two further letters which he wrote to Senators Howard Baker and Albert Gore (from his home state of Tennessee) had an effect. The senators prompted the Atomic Energy Commission to make investigations which confirmed his allegations (Houston, 25).

Was Houston justified in going outside Stone and Webster with his warnings? In retrospect, perhaps it would have been better for him to try working further within the organization first, for example, by writing to division headquarters. Yet there seems little doubt that his actions in pursuing the matter were highly praiseworthy. Because there was so much at stake in the risks of a nuclear plant disaster, the obligation to protect the public had a clear priority in this case.

Moral Guidelines to Whistleblowing

Under what conditions are engineers justified in going outside their organizations when safety is involved? This really involves two questions: When are they morally *permitted,* and when are they morally *obligated,* to do so?

Richard T. De George has suggested that it is morally *permissible* for engineers to engage in external whistleblowing concerning safety when three conditions are met (De George, 6):

1 if the harm that will be done by the product to the public is serious and considerable;

2 if they make their concerns known to their superiors; and

3 if getting no satisfaction from their immediate superiors, they exhaust the channels available within the corporation, including going to the board of directors.

In order for the whistleblowing to be morally *obligatory,* however, De George gives two further conditions (De George, 6):

4 He [or she] must have documented evidence that would convince a reasonable, impartial observer that his [or her] view of the situation is correct and the company policy wrong.

5 There must be strong evidence that making the information public will in fact prevent the threatened serious harm.

De George sets forth these conditions as rough general rules, something like moral rules of thumb. Exceptions and additions can be made. His account allows for the possibility of instances of permissible whistleblowing which do not meet all of conditions 1 through 3. For example, situations of extreme urgency may arise in which there is insufficient time to work through all the normal organizational channels.

We should also add that there may be personal obligations to family and others which militate against whistleblowing. And where blowing the whistle openly could result not only in the loss of one's job but also in being blacklisted within the profession, the sacrifice may in some cases be too much to demand, or may become supererogatory—more than one's basic moral obligations require. Or, what is more likely, anonymous whistleblowing may be the only morally mandatory action.

Nevertheless, conditions 1 through 5 give strong support to the importance of engineers' obligations to the public in areas of safety while at the same time allowing for the importance of their obligations to employers by underscoring the need to work first within organizational channels when trying to correct problematic or dangerous situations. And we agree with De George that when all five conditions are met there arises a very strong prima facie obligation to whistleblow.

George B. Geary and U.S. Steel

We have been focusing upon external whistleblowing, where information is passed outside the organization. Let us now consider a case of internal whistleblowing where the information was conveyed within the organization, although outside regular organizational channels.

George Geary had worked 14 years for the U.S. Steel Corporation. In 1967 he was a sales executive with the company's oil and gas industry supply division in Houston. U.S. Steel was about to market a new type of pipe which Geary believed had been insufficiently tested and might be defective. If the pipe should burst or break while in use, not only would property be damaged, but such a failure might also cause serious injuries to customers and the public.

Geary expressed his strong objections to midlevel management, which decided to go ahead with marketing the new pipe anyway. So, while obeying directives to sell the pipe, he sent his objections to U.S. Steel's higher management. Largely because of his good reputation within the company, top officials took the assertions seriously. They ordered a major reevaluation of the pipe and withdrew it from sales until the tests were completed. Yet shortly thereafter, Geary was fired on the ground of insubordination in going over his manager's head in a matter beyond his area of expertise.

U.S. Steel then attempted to block his unemployment compensation by arguing that he was guilty of willful misconduct. However, the Unemployment Compensation Board of Review, upon hearing the case, reached the following conclusion:

> No company places a man in the position held by the claimant and pays him the salary received by the claimant simply to have him quietly agree to all proposals. The claimant did not refuse to follow orders, but, in fact, agreed to do as instructed despite his opposition to the program proposed. Although he may have been vigorous in his opposition and offended some superiors by going to a vice president, it is clear that at all times the claimant was working in the best interest of the company and that the welfare of the company was primary in his mind. Under these circumstances, giving due regard to the claimant's position with the company, his conduct cannot be deemed willful misconduct (quoted in Nader, 1972, 155–156).

It is possible that Geary's actions prevented injuries, reduced consumer costs, and even saved significant costs to U.S. Steel from premature marketing of a defective product. By all the evidence, he acted as a loyal employee, concerned at once for the good of U.S. Steel and of the public. Yet in communicating information against his immediate superior's wishes, he suffered the same fate as the external whistleblowers discussed above.

Protecting Whistleblowers

The three cases we have examined may seem to present a one-sided, negative picture of what happens to whistleblowers. Whistleblowing does not always have such unfortunate results. Yet most whistleblowers have suffered unhappy, even tragic, fates. In the words of one lawyer who defended a number of them:

> Whistleblowing is lonely, unrewarded, and fraught with peril. It entails a substantial risk of retaliation which is difficult and expensive to challenge. Furthermore, "success" may mean no more than retirement to a job where the bridges are already burned, or monetary compensation that cannot undo damage to a reputation, career and personal relationships (Raven-Hansen, 44).

Yet the vital service to the public provided by many whistleblowers has led increasingly to public awareness of a need to protect them against retaliation by employers. Government employees have won important protections. The Civil Service Reform Act of 1978 protects them against reprisals for lawful disclosures of information believed to show "a violation of any law, rule, or regulation, mismanagement, a gross waste of funds, an abuse of authority, or a substantial and specific danger to public health and safety" (Raven-Hansen, 42). The fact that few disclosures are made appears to be due mostly to a sense of futility—the feeling that no corrective action will be undertaken. Within the nongovernment, private enterprise sector, employees are covered by statutes forbidding firing or harassing of whistleblowers who report to government regulatory agencies the violations of some twenty federal laws, including those covering coal mine safety, control of water and air pollution, disposal of toxic substances, and occupational safety and health. In a few instances unions provide further protection.

Aside from these exceptions, however, most states still allow employers to fire employees they consider "disloyal" at will. There has yet to be full legal recognition of the right of salaried engineers to adhere to professional codes of ethics. But the laws concerning whistleblowing are in transition, and a number of observers believe they are moving in directions favorable to responsible whistleblowing (Walters, 1975, 34; Ewing, 1977, 113; Westin, 1981, 163–164).

We should also add that beyond the protection afforded by law in cases of whistleblowing, there is an important potential role for professional socieites to play. Until recently those societies were reluctant to become involved in supporting engineers who followed the entries in their codes of ethics calling for members to "notify proper authorities" when overruled by their superiors in their professional judgments about dangers to the public. But this is changing. In the BART case, as we saw in Chap. 3, the Institute of Electrical and Electronics Engineers (IEEE) was willing to write a friend-of-the-court brief seeking to establish legal recognition of the right of engineers to act in accordance with professional codes of ethics. IEEE has also established awards and other forms of honorary recognition for whistle-blowers who act according to its ethical code, and furthermore has helped locate new jobs for discharged engineers. Another avenue of protection for engineers being explored by professional societies is the publication in their journals of the names of companies who take unjust reprisals against whistleblowers.

Common Sense Procedures in Whistleblowing

It is clear that a decision to whistleblow, whether within or outside an organization, is a serious matter that deserves careful reflection. And there are several rules of practical advice and common sense which should be heeded before doing so (Unger, 56–57; Westin, 1981, 160–163):

1 Except for extremely rare emergencies, always try working first through normal organizational channels. Get to know both the formal and informal (unwritten) rules for making appeals within the organization.

2 Be prompt in expressing objections. Waiting too long may create the appearance of plotting for your advantage and seeking to embarrass a supervisor.

3 Proceed in a tactful, low-keyed manner. Be considerate of the feelings of others involved. Always keep focused on the issues themselves, avoiding any personal criticisms that might create antagonism and deflect attention from solving those issues.

4 As much as possible, keep supervisors informed of your actions, both through informal discussion and formal memorandums.

5 Be accurate in your observations and claims, and keep formal records documenting relevant events.

6 Consult colleagues for advice—avoid isolation.

7 Before going outside the organization, consult the ethics committee of your professional society.

8 Consult a lawyer concerning potential legal liabilities.

The Right to Whistleblow

Whistleblowers who proceed responsibly and take special care to document their views are fulfilling their obligations to protect and serve the public. To this extent they have a professional moral right to whistleblow. This important right is a restricted one, however, and its appropriate extent can vary according to a number of factors.

Engineers working for the government have as public servants an especially strong charge to protect the public. It is appropriate that they have correspondingly strong rights to whistleblow in the public interest. That right is limited by legitimate needs to keep some information confidential. But not all information stamped "classified" is legitimately confidential. There have been many instances where government corruption has been hidden under claims of confidentiality. The electronic surveillance involved in Watergate is but one example. Moreover, even where legitimate confidentiality is involved, the public interest served by whistleblowing may be of even greater importance.

Engineers working in the private sector also have obligations to the public, especially those based on the right of the public to make informed decisions concerning the use of technological products. Thus they also have a professional right to whistleblow when such action is justified by the appropriate conditions. And we along with the observers noted above, hope and believe that the courts and professional organizations will continue to expand the legal and institutional recognition and protection of this right during the coming decade.

Beyond Whistleblowing

Sometimes whistleblowing is a practical moral necessity. But generally it holds little promise as the best possible solution to remedying problems and should be viewed as a last resort.

The obvious way to remove the need for *internal* whistleblowing is to allow greater freedom and openness of communication within the organization. That is, the need to violate the often rigid channels of communication within organizations would be removed by making those channels more flexible and convenient. But this means more than merely announcing formal "open-door" policies and appeals procedures which give direct access to higher levels of management. Those would be good first steps, and a further step would be the creation of an ombudsperson or an ethics review committee with genuine freedom to investigate complaints and make independent recommendations to top management. The crucial factor which must be involved in any structural change, however, is the creation of an atmosphere of tolerance. There must be a positive affirmation of engineers' efforts to assert and defend their professional judgments in matters involving ethical considerations. Any formal policy can be subverted by supervisors who are preoccupied with their own authority or who create a climate of intimidation. It can also be subverted by those engineers who are insensitive to the legitimate needs of management.

Creating such an atmosphere, then, requires the efforts of management and engineers alike. But it falls on the shoulders of top management to give this aspect of the organization equal priority with other organizational needs and goals. Management's tools include the formal ones of classes and workshops for employees. At Fluor Corporation, for example, a course in engineering ethics is an integral part of an in-house masters degree program in engineering which it finances for its employees. Ultimately, however, it is management's example and style that are decisive in communicating ethical concern to those lower down in the hierarchy of authority. Such concern can be stifled by just one act of inflicting a drastic penalty on an engineer for failing to follow the letter of organizational procedures in zealously pursuing ethical concerns, and it is management's responsibility not to produce such an intimidating atmosphere.

What about *external* whistleblowing? Much of it can also be avoided by the same sorts of intraorganizational modifications. Yet there will always remain troublesome cases where top management and engineers differ in their assessments of a situation even though both sides may be equally concerned to meet their professional obligations to safety.

To date, the assumption has been that management has the final say in any such dispute. But our view is that engineers have a right to some further recourse in seeking to have their views heard.

It is impossible to generalize concerning what this recourse should be within all contexts. Minimally we think it essential that engineers be allowed to discuss—in confidence—their moral concerns with the ethics committees of their professional societies. And it is highly desirable that representatives from those committees, or perhaps professional arbitrators of some sort, be allowed to enter into discussions between engineers and management which have reached a deadlock—again in confidence, as far as the public is concerned. Such was the purpose of Ralph Nader's short-lived Clearinghouse for Professional Responsibility, which sought to serve as a first-step arbiter to resolve employer-employee conflicts in-house.

Beyond this, on-going piecemeal changes within the law, within regulatory bodies, and within corporations themselves must be explored. Some will argue for strong legislation favorable to whistleblowing. But this would allow greater public control over private corporate goals, and management could be expected to resist such outside threats to its autonomy. How far we, as a society, should support laws favorable to whistleblowing will ultimately be as difficult a decision as any other concerning public regulation of private enterprise.

Study Questions

1 Consider the following example:

> Harry works as a designer for a component supplier and often sits in on meetings with clients to keep abreast of their needs. He attends a meeting of corporate executives who decide to phase out a particular component—an encapsulated assembly on which the company is losing money at current production levels. The company will have a new component on the market in six months that not only performs the same functions, but also does additional peripheral functions. It will not be a plug-in replacement for the older component. The client is obviously making a long-term commitment in his design of a system, and the component apparently is key to that system. The client assumes that, in time, when maintenance is required on his system, he will be able to order replacements, if needed. No other company makes a plug-in equivalent of this component. Harry quietly asks the nearest salesman if he is aware that the component is being discontinued, and finds out that the salesmen know its impending fate, but are remaining silent, since the newer component won't be available for six months, and the client is looking to buy something now. If the company doesn't have something to sell him, he'll look elsewhere. What should Harry do? (Perry, 56–57)

In answering this question assume that Harry's immediate supervisor tells him not to do anything at all.

2 Present and defend your view as to whether or not, in the case described below, the actions of Ms. Edgerton and her supervisor were morally permissible, obligatory, or admirable. Did Ms. Edgerton have a professional moral right to act as she did? Was hers a case of legitimate whistleblowing?

  In 1977 Virginia Edgerton was senior information scientist on a project for New York City's Criminal Justice Coordinating Council. The project was to develop a computer system for use by New York District Attorneys in keeping track of data concerning court cases. It was to be added on to another computer system, already in operation, which dispatched police cars in response to emergency calls. Ms. Edgerton, who had thirteen years of data processing experience, judged that adding on the new system might result in overloading the existing system in such a way that the response time for dispatching emergency vehicles might be increased. Because it might risk lives to test the system in operation, she recommended that a study be conducted ahead of time to estimate the likelihood of such overload.

  She made this recommendation to her immediate supervisor, the project director, who refused to follow it. She then sought advice from the Institute of Electrical and Electronics Engineers, of which she was a member. The Institute's Working Group on Ethics and Employment Practices referred her to the Manager of Systems Programming at Columbia University's Computer Center, who verified that she was raising a legitimate issue.

  Next she wrote a formal memo to her supervisor, again requesting the study. When her request was rejected, she sent a revised version of the memo to New

York's Criminal Justice Steering Committee, a part of the organization for which she worked. In doing so she violated the project director's orders that all communications to the Steering Committee be approved by him in advance. The project director promptly fired her for insubordination. Later he stated: "It is . . . imperative that an employee who is in a highly professional capacity, and has the exposure that accompanies a position dealing with top level policy makers, follow expressly given orders and adhere to established policy" (Edgerton Case).

3 It has been suggested that cases of whistleblowing which involve organizational disobedience are similar to instances of civil disobedience (Otten, 182–186; Elliston, "Civil Disobedience and Whistleblowing"). Civil disobedience was a major social tactic used, for example, in the civil rights movement in the 1960s. It may be defined as having the following features: It involves the intentional breaking of a law or government policy; it is nonviolent; it is conducted publicly (rather than secretively); it is performed by generally loyal citizens (as opposed to anarchists and revolutionaries) seeking to change what they believe to be seriously immoral laws or government actions; and participants do not attempt to evade the legal penalties attached to such activity.

    Discuss the similarities and differences you see between civil disobedience and *(a)* the open, external, whistleblowing of Ernest Fitzgerald, *(b)* the open, internal, whistleblowing of George Geary, and *(c)* anonymous whistleblowing.

4 Do you see any special moral issues raised by *anonymous* whistleblowing? For a helpful discussion consult Frederick Elliston's essay, "Anonymous Whistleblowing" (Elliston).

EMPLOYEE RIGHTS

Employee rights are any rights (moral or legal) which have to do with the status of being an employee. They include some professional rights which apply to the employer-employee relationship: for example, the right to disobey unethical directives and to express dissent from company policies without employer retaliation. Thus the professional rights discussed in the previous section are also employee rights insofar as they relate to the condition of being a salaried professional.

    Then too, employee rights include fundamental human rights relevant to the employment situation. In the next section we will discuss one of those: the right not to be discriminated against because of one's race, sex, age, or national origin.

    And one group of employee rights are institutional rights created by organizational policies or contracts. For example, an engineer whose negotiated salary is $30,000 has a contractual right to that amount of money. He or she may also have contractual rights to various company benefits, such as periodic pay raises and profit sharing. These rights are based solely on employment contracts.

    However, a different group of employee rights will be the topic of this section. In contrast with purely contractual rights, these exist even if unrecognized by specific contract arrangements or company policies. Companies and employers *ought* to recognize them, whether or not they actually do. For they are more than mere privileges which employers are permitted to disregard.

Ewing's Employee Bill of Rights

In *Freedom Inside the Organization,* David Ewing, editor of *The Harvard Business Review,* refers to employee rights as the "black hole in American rights." The Bill of Rights in the Constitution was written to apply to government, not to business. But when the Constitution was written no one envisaged the giant corporations which have emerged in our century. Ewing demonstrates compelling parallels between the kinds of threats to liberty posed by large and powerful governments (which the Founding Fathers wrote the Constitution to protect citizens against) and the kinds of threats to individual freedom posed by present-day business organizations. Corporations wield enormous power politically and socially, and especially over their employees. They operate much as minigovernments, and are often comparable in size to those governments the Founding Fathers had in mind when they wrote the Constitution. For instance, American Telephone & Telegraph in the 1970s employed twice the number of people inhabiting the largest of the original thirteen colonies when the Constitution was written.

Ewing proposes that large corporations ought to recognize a basic set of employee rights. He gives the following concise statement of what those rights should involve:

> No public or private organization shall discriminate against an employee for criticizing the ethical, moral, or legal policies and practices of the organization; nor shall any organization discriminate against an employee for engaging in outside activities of his or her choice, or for objecting to a directive that violates common norms of morality.
>
> No organization shall deprive an employee of the enjoyment of reasonable privacy in his or her place of work, and no personal information about employees shall be collected or kept other than that necessary to manage the organization efficiently and to meet legal requirements.
>
> No employee of a public or private organization who alleges in good faith that his or her rights have been violated shall be discharged or penalized without a fair hearing in the employer organization (Ewing, 1977, 234–235).

In previous sections we discussed some of these rights, such as the rights to free speech and dissent and the right of conscientious refusal to obey unethical directives. Here we will examine several others—in particular, those relating to the choice of outside activities, to privacy, and to due process.

Freedom to Choose Outside Activities

All employees have the right to pursue nonwork activities of their own choosing without coercion or retribution from employers. This is part of their basic human right to pursue legitimate interests without interference. But because this right has generally not been protected by state or federal laws, there have been some flagrant violations of it.

For example, a worker in a Ford Motor Company service department was fired because his supervisor learned he had bought a new American Motors Rambler instead of a Ford automobile. Because he happened to be a union member, he was able to regain his job. Others have not been as lucky and have had to buckle to pressures from employers. Or there is the case of an executive for Phillips Petroleum who stated in a national interview that he did not want to see Phillips employees at competitors' gas stations (Ewing, 1977, 120–121).

Such abuses are perhaps becoming rarer, especially in states like California and Florida which have passed laws prohibiting them. But why, we might ask, would employers make such demands and intrusions into the personal purchasing habits of their employees?

No doubt part of the answer lies in an exaggerated concern for company loyalty. Loyalty comes to be viewed as extending beyond the fulfilling of job functions into areas of personal decision making. A more important part of the answer, however, is the extreme concern companies have to present a unified and untarnished image to the public. Even the slightest or most indirect damage to that image, and to the employers' ability to control the image, is perceived as a threat. One such threat lies in the negative attitudes toward the company that could potentially arise when it is learned by outsiders that employees are not purchasing their own company's products. Yet an employer's rightful concern with the company image should not extend to control over employees' personal buying habits.

Consider a different example. In 1971 IBM fired Lawrence Tate, an engineer employed by that company for over 18 years. Tate had a shining record with IBM and was fired solely because of a certain article that appeared in a local newspaper. The article told of Tate's efforts to reform the police department in his area. It also mentioned his subsequent arrest and conviction on two misdemeanor charges and his daughter's arrest for possession of one marijuana cigarette. The convictions were being appealed, and he was countercharging police harassment. Unfortunately, the article also mentioned Tate's place of employment, and it was this which led to his being fired (D. Fitzgerald, 197–198).

Perhaps some indirect damage to IBM's reputation resulted from the article. But if this sort of thing can be a legitimate reason for firing an employee, none of us has much genuine freedom. The single act of a careless journalist would be enough to undermine our careers.

What about a corporation's right to protect its public image? Surely this places some limits on the rights of employees to pursue outside activities, in spite of what has been said so far. No simple line can be drawn here, but a few generalizations are possible:

First, the rights of employees to pursue outside activities become limited at the point where those activities lead to violations of the duties connected with their jobs. Here what is actually at stake are not the outside activities per se, but their effects on job-related activities. For example, an individual has the right to abuse alcohol without interference in the privacy of his or her own home even though such conduct may be foolhardy and some people upon learning of it may lower their estimate of the company the person works for. But the employer has a right to take action against the employee if the alcohol abuse begins to damage work performance.

Second, employers have the right to take action when outside activities constitute a conflict of interest. Here there may be no actual harm done, as we saw in Chap. 5. But the potential exists for failure to fulfill duties to the organization and for fostering a public image of tenuous employee loyalty to the company. Employers are plainly within their rights, for example, in requiring a person to stop moonlighting for a competitor's business.

Third, employees have no right to consistently sabotage their employers' interests during off-hours. During labor disputes, a mutual recognition of each other's legitimate interests must be maintained.

In every case, however, we should add, the burden of proof should always fall on the employer to establish that the corporation's interests are so compelling as to take precedence over the rights of its employees.

Right to Privacy

The right to pursue outside activities can be thought of as a right to personal privacy in the sense that it means the right to have a private life off the job. In speaking of the right to privacy here, however, we mean the right to control access to and use of information about oneself. (As with the right to outside activities, this right is limited in certain instances by employers' rights.)

Consider a few examples where the functions of employers come into conflict with the right employees have to privacy:

1 Before being hired at a computer center which handles large banking transactions, applicants are required to answer questions about their past criminal records while taking a polygraph (lie detector) test.

2 Job applicants at the sales division of an electronics firm are required to take personality tests which include personal questions about alcohol use and sexual conduct. The rationale given for asking those questions is a sociological study showing correlations between sales ability and certain

data obtained from answers to the questionnaire. (That study has been criticized by other sociologists.)

3 A supervisor unlocks and searches the desk of an engineer who is away on vacation without the permission of that engineer. The supervisor suspects the engineer of having leaked information about company plans to a competitor and is searching for evidence to prove those suspicions.

4 A sociologist has been hired as a consultant to a large construction firm which has been having personnel conflicts in one division. Without checking with its employees, management gives the sociologist full access to its personnel files.

5 A large manufacturer of expensive pocket computers has suffered substantial losses from employee theft. It is believed that more than one employee is involved. Without notifying employees, hidden surveillance cameras are installed.

6 A rubber-products firm has successfully resisted various attempts by a union to organize its workers. It is always one step ahead of the union's strategies, in part because it monitors the phone calls of employees who are union sympathizers. It also pays selected employees bonuses in exchange for their attending union meetings and reporting on information gathered. It considered, but rejected as imprudent, the possibility of bugging the rest areas where employees were likely to discuss proposals made by union organizers.

Some of these examples involve abuse of employer prerogatives. Most of them involve a clash between the right to privacy of employees and the right of employers to effectively manage a corporation. We may differ in our opinions about some of them. But surely such intrusions are morally problematic and stand in need of special justification. Why is it that privacy is so important, even at work, and what is the basis of the right to it?

In order to answer that question it will be useful to postulate an extreme, hypothetical case (Fried, 138). Imagine a scenario akin to that described in George Orwell's *1984* which is feasible with current technology: Unknown to them, all employees in a firm—engineers, skilled laborers, secretaries—have a tiny electronic device implanted in their plastic identification badges. The device emits steady signals indicating the location of the employee carrying it and can be used by a distant monitoring center to record conversations taking place within a radius of 20 feet around it. Why would this situation upset us if we were the employees?

One answer is that we would object to the deception involved. So to focus matters directly on privacy, let us imagine that we are fully informed of the use and nature of the bugging device. We are not asked to give our consent, however, but are told that allowing the device to be used is a condition of our employment. Why would we still object?

A utilitarian philosopher might answer that such a situation would make us unhappy for various reasons. It would lead to a general apprehensiveness about how our every word might be interpreted by those who could hear our conversations. Thus our conversational spontaneity would be inhibited. Certainly the use of such a device would destroy any sense of being trusted by our employer. And it would open the door to innumerable abuses, such as harassment, by those who might have access to what the device recorded or picked up.

A duty-based ethicist might argue that use of the device would violate the duty to respect people. Respect for people entails allowing them some degree of control over who has knowledge about their personal conversations (Benn, 8–9, Reiman, 387). It also means the duty to allow others the pursuit of intimate relationships—friendships, trust-relationships, etc.—would be harmed by the use of such devices (Fried, 137–152). Intimacy involves selectively revealing information about ourselves which we otherwise keep secret, and revealing it in the belief that it will not be used to harm us. We would not reveal as much if we knew others were listening. And in the work environment, a climate of fear would prevent even the normal jokes between colleagues about bosses which help contribute to comradeship.

Finally, a rights-based ethicist would appeal directly to the human right of personal freedom: People should be free to maintain some control over what personal information about themselves is revealed to others. Denying anyone this freedom destroys a rich dimension of choice in expressing oneself and developing personal relationships.

The right to privacy is limited by the legitimate exercise of employers' rights to obtain and use information necessary for the effective managing of an organization. These rights make it legitimate to employ aptitude and skill tests related to job functions. But they do not make it legitimate to use general personality and intelligence tests which have not been established as essential for measuring job performance.

Once gathered, information about employees should be reserved solely for legitimate employer use. It is not permissible to give it to outsiders, or even to members of the corporation who do not need to know it. The personnel division, for example, needs medical and life insurance information about employees, but immediate supervisors usually do not. Only the most routine information about job position and years with the company should be revealed to inquirers from outside the corporation, such as collecting agencies, insurance companies, apartment landlords, and private investigators. Such protective procedures can be simply and effectively implemented. IBM led the way years ago in voluntarily initiating such a program which is serving as a model for other corporations (Ewing, 1976, 82).

Employers should be viewed as having the same fiduciary or trust relationship to their employees concerning confidentiality that doctors have to their patients and lawyers have to their clients (Mironi, 289). In all of these cases personal information is given in trust on the basis of a special professional relationship. Moreover, with rare exceptions involving identification of other parties, employers owe employees the right to examine their dossiers so as to correct outdated or erroneous information.

Due Process

Rights of conscience, free speech, outside activities, and privacy would be of little help to employees were they not given institutional recognition. Institutional recognition includes formal endorsement of the rights, such as in company policy statements or employment contracts. But it also requires creating an institutional procedure for protecting those who exercise the rights. Thus, the substantial rights discussed so far imply a right to due process—that is, a right to fair procedures safeguarding the exercise of other rights. The right to due process extends to fair procedures in firing, demotion, and disciplinary actions.

Implementing the right to due process involves two general procedures: First, written explanations are owed to employees who are discharged, demoted, transferred to less enriching work, or in other ways penalized.

Second, an appeals procedure should be established which is available to all employees who believe their rights have been violated. The procedure should be a stable part of the organization, effective, equitable, and efficient. For the sake of both management and employees, it must be easy to use and work quickly, generally yielding a verdict within days after a grievance is filed (Ewing, 1977, 155–174).

Government employees and union members generally have some such procedures available to them, however flawed they may be in practice. Private companies have recently developed a variety of promising procedures, some of which are still largely experimental. Polaroid, for example, has set up a grievance committee composed of members elected by employees. Xerox has a formal employee advisory board. General Electric uses an impartial referee. Some corporations have ombudspersons who hear and investigate complaints.

The power of these various appeals groups and people is limited, but it can be substantial. It is limited because it typically involves making recommendations to top management, without issuing final verdicts binding on management. Binding verdicts, by definition, would remove substantial decision-making authority from management. The ability to issue them would in effect create a new line of management authority. Yet the influence of appeals bodies can be significant: Where employees trust them, overriding their decisions without very compelling reasons can cause serious personnel problems.

Study Questions

1 Early in 1970 the proposal was made to prohibit all non-American-made cars from parking inside the gates of a major steel production facility. The proposal grew out of the negative impact imported steel products were having on the domestic American steel industry. The rule was to apply to all employees, including management (Barry, 164–165).

    Present and defend your view as to (a) whether this proposal should have been approved and (b) whether such issues should be decided by a majority vote among employees or in some other way.

2 In 1974 Combustion Engineering required its officers and employees to take lie detector tests. Someone had given confidential information concerning the terms of its nuclear power contracts to *The Wall Street Journal,* and management felt justified in finding out who it was (Ewing, 1977, 131). Explain your view as to whether or not this tactic was justified, or whether it might have been justified depending on further details of the case.

3 Explain and defend your views concerning examples 1 through 6 in the above section on the right to privacy. Was management in each case justified in doing what it did? Can you formulate any general guidelines on the basis of the examples?

4 Some observers have argued that employees have a right to choose the type of clothing and hair styles they wear to work. The courts, however, have ruled that employers have the right to set reasonable standards, specifically in instances where corporate image and job function are affected. Discuss this issue, defending your own view by reference to utilitarian, duty-based, or rights-based ethical theories.

5 The chairman of the board of directors of a company whose main business comes from government military contracts sends a letter to all employees. The letter, written on company stationary, outlines why it is clearly in the interests of the company that a certain promilitary senator be elected. The letter concludes by stating that while no employee is required to make a campaign donation to the senator, those wishing to do so can have the donation deducted from their next paycheck. Is there anything objectionable about this? Are any rights infringed?

6 Consider the following example:

Several top executives of a company in a large city are disturbed by community activist organizations that are protesting the treatment of minority groups. The executives feel that the activists, while abstaining from violence, are doing more harm than good to the schools, urban renewal programs, public transportation, and retail business. The chief executive himself has articulated his fears about the activists at local business meetings. However, a young . . . [engineer working for the company] thinks that the activists are on the right track. He goes to work for the leading activist organization, spending many evening and weekend hours doing unpaid volunteer tasks. Occasionally he is quoted in the newspaper and identified with the employer company (Ewing, "What Business Thinks about Employee Rights," 31).

Does the engineer have a right to continue his activities without interference from the employer? Or do the executives have a right to tell him to stop the activities or else be fired?

DISCRIMINATION

Perhaps nothing is more demeaning than to be downgraded for one's sex, race, skin color, age, or religious outlook. These aspects of biological makeup and basic conviction lie at the heart of self-identity and self-respect. Such downgrading is especially pernicious within the work environment, for work is itself fundamental to a person's self-image. Accordingly, human rights to fair and decent treatment at the workplace and in job training are vitally important.

Yet there are challenging moral issues concerning those rights. One concerns the role of government. Should government and the law be used within private enterprise to oppose discrimination? Or do corporations have the right to hire whomever they please in the search for profits and economic efficiency?

A second set of issues concerns the appropriate extent of the right to nondiscrimination. For example, do women and minorities who in the past have been discriminated against have the right to be given preferential treatment in entrance to educational programs, in hiring, and in job retention? Or does such preference violate the right to equal opportunity enjoyed by members of majority groups and thus itself constitute discrimination "in reverse?"

These issues are of significance to engineers. Traditionally, engineering in the United States has been one of the more open avenues for upward mobility of capable but otherwise disadvantaged persons. Partly this has to do with the willingness of engineers to recognize talent where they see it, partly it is the result of government-mandated fair-employment practices which must be observed by employers engaged in government contracts. But in times of economic downturns the influx of women, racial minorities, and foreign nationals into the engineering workforce can produce unfavorable reactions. Should there be any conflicts, engineers must carefully examine the ethical bases of their and their colleagues' actions.

Examples

Consider the following examples:

1 An opening arises for a chemical plant manager. Normally such positions are filled by promotions from within the plant. The best qualified person in terms of training and years of experience is a black engineer. Management believes, however, that the majority of workers in the plant would be disgruntled by the appointment of a nonwhite manager. They fear lessened employee cooperation and efficiency. It is decided to promote and transfer a white engineer from another plant to fill the position.

2 Several women engineers work in the sales division of an electronics company. The company prides itself on the practice of hiring proportional numbers of women. Yet the pay scale for these women is systematically lower than that for men having comparable experience and engaging in comparable work. In the absence of objections, management assumes that the women recognize the necessity to pay males more because they are the primary breadwinners for their families.

3 A farm-equipment manufacturer has been hit hard by lowered sales caused by a flagging produce economy. Layoffs are inevitable. During several clandestine management meetings, it is decided to use the occasion to "weed out" some of the engineers within 10 years of retirement in order to avoid payments of unvested pension funds.

Definition

The word "discrimination" is used in several senses. Sometimes it means preference on the grounds of sex, race, etc., whether or not such preference is viewed as justified. In everyday speech, however, it has come to mean *morally unjustified treatment of people on arbitrary or irrelevant grounds.* We will use the word in this latter sense. Thus to call something "discrimination" is to condemn it. Where the question of justification is left open for discussion, we will speak of "preferential treatment."

Most of us would agree that the preceding examples involve discrimination. They also involve violation of antidiscrimination laws. Let us review some of these laws before inquiring into whether they are morally warranted.

Antidiscrimination Laws

The forerunner of antidiscrimination laws was the principle of equality which the Fourteenth Amendment embedded in the Constitution following the Civil War (1868):

No State shall make or enforce any law which shall abridge the privileges or immunities of citizens of the United States; nor shall any State deprive any person of life, liberty, or property, without due process of law; nor deny to any person within its jurisdiction the equal protection of the laws.

But it was only as recently as 1964 that discrimination by public or private employers was explicitly prohibited legally in the Civil Rights Act:

It shall be an unlawful employment practice for an employer to fail or refuse to hire or to discharge any individual, or otherwise to discriminate against any individual with respect to his compensation, terms, conditions, or privileges of employment, because of such individual's race, color, religion, sex, or national origin (Title VII—Equal Employment Opportunity).

The Equal Employment Opportunity Act of 1972 amended and strengthened the Civil Rights Act by giving greater powers of enforcement to the Equal Employment Opportunity Commission.

Several supporting Executive Orders, having the force of law, were issued under Presidents Kennedy and Johnson. They required that businesses receiving government contracts develop Affirmative Action Programs to remedy underrepresentation of women and minorities on their staffs. "Underrepresentation" usually meant that the percentage of women and minority workers hired for a given type of job did not roughly parallel the percentage of those available locally to be hired.

As originally mandated, Affirmative Action programs had a dual emphasis. On the one hand, they sought to increase the number of women and minority applicants for jobs and professional education. This was to be accomplished largely through wide advertising of the positions and opportunities involved, assuring applicants that they would be considered on a nondiscriminatory basis. On the other hand, the programs called for setting concrete goals for hiring, including timetables for achieving racial or sexual job parity. Explicit numerical quotas, however, were not endorsed. Reaching those goals usually meant hiring women and minorities over equally qualified white males. In practice, government pressure also sometimes led to giving them preference over better-qualified white males, although this was not justified by the law.

Finally, age discrimination was prohibited by the Age Discrimination in Employment Act of 1967, amended in 1974. That act states:

Sec. 4. (a) It shall be unlawful for an employer—(1) to fail or refuse to hire or to discharge any individual or otherwise discriminate against any individual with respect to his compensation, terms, conditions, or privileges of employment, because of such individual's age;
(2) to limit, segregate, or classify his employees in any way which would deprive or tend to deprive any individual of employment opportunities or otherwise

adversely affect his status as an employee, because of such individual's age (quoted in Sethi, 334).

Moral Justification of Nondiscrimination Laws

The equal opportunity laws forbid the kinds of discimination involved in the examples given earlier. But are they morally justified? Should the government be allowed to use law to force private corporations to treat people equally?

The libertarian (or conservative free market) position answers these questions in the negative. It argues that government should not meddle in private economic dealings except where necessary to protect contracts and free competition. According to this view, all fair employment legislation is unjust. As Milton Friedman wrote in *Capitalism and Freedom:* "Such legislation clearly involves interference with the freedom of individuals to enter into voluntary contracts with one another" (Friedman, 1962, 111). Forcing a grocer to hire a black clerk in a neighborhood which hates blacks is, he says, unfairly hurting the grocer's business. In general, fair employment legislation lessens economic efficiency and violates the right of businesses to hire whomever they please.

Friedman adds that he personally finds racism and sexism deplorable. But he regards this merely as his personal preference, and insists that neither he nor the majority has a right to force their preferences on others. "It is hard," he writes, "to see that discrimination can have any meaning other than a 'taste' of others that ones does not share" (Friedman, 1962, 110).

Friedman denies there are moral rights bearing on employment which are not reducible to contracts and to the conditions which make contracts possible. In our view this is a serious mistake. It is not a mere personal preference to condemn racist and sexist practices. Laws forbidding prejudicial employment practices are not a matter of a majority imposing its tastes on others. For those laws protect the basic moral rights everyone possesses to have a fair opportunity of working to obtain social benefits.

Those rights are not absolute in the sense that nothing could ever override them. Presumably if utter economic disaster resulted from recognizing them, they might have to be somewhat restricted for a time. But it is simply not true that fair employment legislation has drastically disrupted the economy. The laws have placed some new restrictions on the exercise of the economic right to pursue profits. But the vital importance of being able to pursue work and careers without crippling interference from discriminatory practices seems a sufficient reason to justify such restrictions.

Let us now turn to the issues surrounding reverse preferential treatment. These are more troublesome because at least on the surface preferential treatment is a violation of the very concept of equality used to condemn past racist and sexist practices.

Reverse Discrimination

Hiring a woman or a member of a minority over an equally qualified white male is only one form of reverse preferential treatment. Let us call it the *weak* form. The *strong* form, by contrast, consists in giving preference to women or minorities over better-qualified white males. The strong version, of course, is the more highly controversial one. It is one thing to give preference by tipping an equally balanced scale. It is quite another to load the scale from the outset. The following discussion is focused upon strong preferential treatment, as are most current debates over "reverse discrimination."

Reverse discrimination, as we shall define it, occurs when unfair preference is given to a member of a group which in the past has been the object of discrimination. Typically this occurs when unfair preference is given to a woman or a member of a previously downgraded minority at the expense of a white male. The question is whether or not it is unfair or otherwise unjust to give preference to some lesser-qualified women or minorities.

The Bakke Decision   Reverse discrimination in regard to education received national attention in the Supreme Court's 1978 Bakke decision. Allan Bakke, a white engineer, was twice denied admission to the medical school at the University of California, Davis. His grades and scores on entrance exams were significantly higher than most of the minority students admitted under a special admissions program. Of 100 openings, the program reserved sixteen for Blacks, Chicanos, Asians, and American Indians. Minority applicants for those sixteen positions were judged only among themselves, and not by comparison with applicants for the other, unreserved openings. Bakke went to court, charging that he had been unfairly discriminated against in violation of the Civil Rights Act.

U.C. Davis attempted to defend its special admissions program on four grounds: (1) it increased the number of minorities in the medical profession; (2) it countered the bad effects of past racial discrimination; (3) it increased the number of physicians likely to serve in inner cities and other areas underserved medically; and (4) it created a mixed student body, which in turn widened the educational benefits available to students thus exposed to a greater intellectual and social diversity.

In a split vote (5 to 4), the Supreme Court ruled that the first three grounds did not justify treating candidates differently because of race. It also ruled, however, that the last ground—the educational importance of a mixed student body—did justify counting race as one relevant consideration in screening candidates.

Nevertheless, it agreed that the U.C. Davis quota system unfairly discriminated against Bakke and other whites. For by reserving certain positions for minorities, the Court argued, it prevented complete comparisons being made among all applicants. Thus Bakke won his case and was allowed to enter the medical school (and has since then been graduated).

The Weber Decision   One year after the Bakke case, the Supreme Court made a ruling which to some people has seemed incompatible with both the Civil Rights Act and the Bakke condemnation of quota systems. The second case concerned a job training program at the Gramercy, Louisiana, plant of Kaiser Aluminum and Chemical Corporation. The area surrounding Gramercy has a 40 percent black population. Yet only a small number of skilled crafts workers at the Kaiser plant were black. This was due to the unavailability of skilled black workers, not to any policy of discrimination at the Kaiser plant.

In 1974 Kaiser and the United Steelworkers of America entered into a collective bargaining agreement to give preference to black applicants for job training programs. Half the positions in these programs were to be reserved for black workers. In addition, a previous eligibility requirement of work experience was abolished. Seniority counted for admission to the programs, but two separate seniority lists were formed, one for black and one for white workers. This policy of preferential treatment was to continue until the percentage of skilled black workers employed by the plant was roughly equal to 40 percent of the skilled workforce.

That same year Brian Weber, a white worker, was denied entrance to one of the programs. Weber had worked at the plant since 1969 and had more seniority than some of the black workers admitted. He filed a class action suit against Kaiser, alleging a violation of Title VII of the Civil Rights Act.

A literal reading of the Civil Rights Act would condemn the Kaiser job training program:

> It shall be an unlawful employment practice for any employer, labor organization, or joint labor-management committee controlling apprenticeship or other training or retraining, *including on-the-job training programs,* to discriminate against any individual because of his race, color, religion, sex, or national origin in admission to, or employment in, any program established to provide apprenticeship or other training [42 U.S. Codes 2000e-2 (d) (1970), emphasis added; quoted in C. Cohen, 375].

Yet the Supreme Court ruled that the literal reading was not the correct one. The justices wrote, "It is a 'familiar rule, that a thing may be within the letter of the statute and yet not within the statute, because not within its spirit, nor within the intention of its makers.' " They contended that the intent of both the Civil Rights Act and the Kaiser program was to eliminate traditional patterns of racial segregation and inequality, and on that basis ruled against Weber (443 U.S. 193 [1979]).

Let us now sketch a few of the main arguments used to justify strong preferential treatment of women and minorities, then some used to prove that it is unjustified and should be considered in effect "reverse discrimination."

Arguments For   A rights-based argument favoring strong preferential treatment emphasizes the principle of compensatory justice: Past violations of rights must be compensated where possible "in kind." Taking property from others minimally requires returning it. Similarly, members of groups who have suffered job discrimination in the past are owed special advantages in obtaining jobs today. Ideally such compensation should be given to *individuals* who in the past were denied jobs. But the costs and practical difficulties of determining such discrimination on a case-by-case basis through the job-interviewing process and the legal system forces a more blanket approach. Preference, therefore, is to be given on the basis of membership in a group which has been disadvantaged in the past.

Those utilitarians who favor preferential treatment have additional arguments. They point to the importance of integrating women and minorities into the economic and social mainstream. Only thus can the benefits of harmony among the races and sexes be achieved. And only thus can our society benefit from the resources of individuals whose potential has been negated in the past. Given the central role work plays in building self-identity and self-esteem, in allowing people to develop their resources to the greatest extent possible, and in providing role models, therefore, preferential treatment in job hiring is seen as the best way to attain these goals.

Arguments Against   There are also forceful arguments against strong preferential treatment. Such preference, it can be argued, is a straightforward violation of other people's rights to equal opportunity. Just as the rights of minorities and women to such equality were regularly violated prior to enactment of the Civil Rights Act, the rights of white males are now being violated whenever their job qualifications are disregarded because of their race and sex. In general, the argument goes, whatever made discrimination against disadvantaged groups unfair in the past also makes such "reverse discrimination" unfair in the present.

Opponents of preferential treatment for minorities often grant that past violations of rights may call for compensation. But they view general policies of hiring the disadvantaged at the expense of equally or even better-qualified white males as an improper way of providing that compensation. Most white males have not themselves participated in discriminatory job actions, and to ignore their rights to fair employment opportunity amounts to compensating victims by punishing the innocent. Two wrongs cannot make a right.

Thus blanket compensation to all members of a group, it is argued, should be given at most in the form of special early education and social programs designed to provide the necessary training to yield a fair opportunity later in applying for jobs or professional schools. Even here, however, economically deprived whites should not be excluded from participation.

Utilitarians who are against strong preferential treatment have additional arguments. They say that the harm such policies do goes beyond that involved in violating the job rights of white males. For example, there is the intense resentment generated among white males and their families. Those feelings only intensify racial tensions and ultimately work against the goals of integration. Moreover, preferential treatment subtly but insidiously encourages traditional stereotypes: A sense of inferiority may arise in women and minorities who come to feel they cannot make it on their own without special help. Finally, there is the economic harm that results from a policy of not consistently hiring the best qualified person.

Intermediate Positions   Recently various attempts have been made to develop intermediate positions sensitive to all the above arguments for and against strong reverse preferential treatment. Two of these will be mentioned here.

The first rejects all blanket preferential treatment of special groups as inherently unjust and a violation of the right to equal treatment of other groups not so treated. Hiring by competence is seen as vital to the well-being of society. Yet the principle of compensatory justice as applied to individuals is also vitally important. If it can be shown that a given company has discriminated against specific individuals, then that company is obligated—and should be required—to give the next available jobs to those individuals. It is true that this might involve discrimination against other, perhaps better-qualified, applicants for those jobs. But the principle of compensatory justice, so the argument goes, in effect automatically closes new jobs to any further applicants beyond the individuals previously discriminated against (Goldman, 1979, 120–127).

In contrast, the second view seeks to justify preferential treatment of special groups. It contends that racist and sexist attitudes are still widespread, at least at a visceral or gut level. Mere affirmative action programs are not sufficient to counterbalance the subtle impact of these attitudes on employment practices. The only adequate way to provide such a counterbalance is to allow and encourage strong preferential treatment. Admittedly some violations of other people's rights to equal employment opportunity will occur in the process, at times with tragic results. But that is the necessary evil we must live with to remedy the deeper tragedy resulting from racism and sexism (Beauchamp, 625–635).

These two intermediate views, together with the preceding pro and con arguments, make it clear that the issues surrounding strong preferential treatment are subtle and complex. Compelling arguments exist on both sides. In resolving the issue, rights-based ethicists must identify a reasoned perspective from which to weigh rights to equal treatment against rights to compensation. Duty-based ethicists must balance duties to treat people equally with duties to provide compensation for past wrongs. Utilitarians must struggle to find a proper way to sum up the positive and negative consequences that can result from such policies. And since nearly all of us will at some time or other be involved in a situation involving preferential treatment, each of us individually needs to find a way to balance these considerations in a carefully reasoned manner.

Sexual Harassment

Sexual harassment is a particularly invidious form of sex discrimination, involving as it does the abuse of not only gender roles and work-related power relationships, but the abuse of sexual intimacy itself. The following discussion focuses on the most widespread type of sexual harassment: male harassment of females. And in regard to the field of engineering, the female may be an engineer, a technician, or a secretary, and the male may be, for example, an engineer-manager or an engineer-colleague.

The term "sexual harassment" is currently applied to a wide variety of physical and psychological attacks, coercion, and sexual practices. One definition of it as applied to women is: "any sexually oriented practice that endangers a woman's job—that undermines her job performance and threatens her economic livelihood" (quoted by Backhouse, 32). Another definition is: "the unwanted imposition of sexual requirements in the context of a relationship of unequal power" (MacKinnon, 1).

Sexual harassment may come in many forms: (1) Following an interview for a job as secretary, a woman is told that the job is hers if she is willing to grant sexual favors to the interviewer. (2) A woman is told by her superior that she will have first priority for receiving a promotion if she is "nice" to him, and talk of a motel makes it clear what is meant by the term "nice." When she refuses to be that "nice," she is not given the promotion and thereafter is assigned less challenging work. (3) Against her will, a woman is grabbed and kissed by her employer, who had asked her to stay after hours at work. She resists and is fired the following day. (4) A woman turns down her boss's request for a date. She makes it clear she is not interested in going out with him ever, but to her chagrin he continues repeatedly to ask her out during the following weeks. (5) The male colleagues of a woman continually leer at her and make sexually suggestive comments about her clothing and body. (6) A male engineer enjoys telling his secretary about his sex life, disregarding her protests against hearing about it.

Although there is evidence that sexual harassment is a widespread and serious problem, the courts have only recently begun to take action. Since 1976 there have been a series of rulings which recognize at least two forms of sexual harassment as instances of sex discrimination prohibited by the Civil Rights Act: first, where a supervisor requires sexual favors as a condition for some employment benefit (a job, promotion, raise); second, where actual job-related retaliation is taken by an employer or supervisor when a sexual request is refused. However, a third, more prevalent, form of sexual harassment has yet to be given serious attention in court rulings: that is, where the harassment functions as part of the everyday work environment among coworkers—for example, situations where women must put up with repeated, unwanted sexual proposals or lewd comments (MacKinnon, 57–82).

What is morally objectionable about sexual harassment at the workplace? Any answer must take into account the generally inferior economic status of women. Such harassment takes advantage of that condition, and it is no surprise that it is directed mostly toward secretaries, clerical workers, and other low-paid female workers. As with rape, sexual harassment is a display of power and aggression through sexual means. Accordingly it has appropriately been called "dominance eroticized" (MacKinnon, 162).

Insofar as it involves coercion, sexual harassment constitutes an infringement of one's autonomy to make free decisions concerning one's body. But whether or not coercion and manipulation are used, it is an assault on the victim's dignity. In abusing sexuality, such harassment degrades people on the basis of a biological and social trait central to their sense of personhood.

Thus a duty-based ethicist like Kant would condemn it as violating the duty to treat people with respect, to treat them as ends-in-themselves and not merely as means to personal aggrandizement and gratification of one's sexual and power interests. A rights-based ethicist would see it as a serious violation of the human right to pursue one's work free from the pressures, fears, penalties, and insults that typically accompany sexual harassment. And a utilitarian would emphasize the impact it has on the victim's happiness and self-fulfillment.

Study Questions

1 Present and defend your view as to (a) whether or not weak preferential treatment of minorities is ever justified; (b) whether or not strong preferential treatment of

minorities is ever justified; (c) whether or not government intervention in the form of enforcing laws like the Civil Rights Act is morally justified.

2 In the early 1960s Motorola screened some 20,000 job applicants each year. In order to increase efficient hiring and decrease the costs of screening, they administered a 5-minute technical test. The Illinois Fair Employment Practices Commission charged that the test was discriminatory against black applicants. Consider the following facts known about the test: (a) The same test was used for both whites and blacks, but the score considered passing was based upon an original test-standardization-group which was predominantly white. (b) There was a proven correlation between general technical ability and high scores on the test. (c) While the test was felt to be a generally reliable indicator for technical trainability, it was known that at least some qualified applicants had been ruled out on the basis of their scores after taking it. (d) There was no evidence either way concerning whether job performance of blacks was more erroneously predicted by results of the test than job performance of whites (Garrett, 47–50).

Would any of these facts show the test to be unfairly discriminatory? What further facts would be relevant in determining the answer to this question?

3 While engineering remains one of the most male-dominated professions, strong efforts have been made to encourage women to enter it. About 14 percent of entering engineering classes are composed of women. The salaries of beginning women engineers are on the average slightly higher than those of men (Vetter, 29–30). Yet women engineers are often subjected to greater pressures than males, both in school and on the job (Davies, 32). Discuss why you think that is so and what might be done about it.

4 Imagine two applicants for a construction supervisor's position. One is a 55-year-old white male engineer. The second is a 30-year-old white male engineer. Both have sufficient professional credentials for the job, but the younger man has fewer years of work experience. The 30-year-old man is hired.

Redescribe the example more fully (without contradicting the given information) in a way that would make it evident that unfair age discrimination was involved in the hiring decision. Then embellish on the example once more (again without contradicting the facts as given) in a way which would indicate that the correct choice was made in a nondiscriminatory way.

5 A company advertises for an engineer to fill a management position. Among the employees the new manager is to supervise is a woman engineer, Ms. X, who was told by her former boss that she would soon be assigned tasks with increased responsibility. The prime candidate for the manager's position is Mr. Y, a recent immigrant from a country known for its confining roles for women. Ms. X was alerted by other women engineers to expect unchallenging, trivial assignments from a supervisor with Mr. Y's background. Is there anything she can and should do? Would it be ethical for her to try to forestall the appointment of Mr. Y?

SUMMARY

Salaried engineers have several overlapping types of moral rights:

I Human rights—possessed by virtue of being people or moral agents
    Examples:

Fundamental right to pursue legitimate interests

Right to make a living

II Professional rights—possessed by virtue of being professionals having special moral responsibilities

Examples:

Basic right of professional conscience (the right to exercise professional judgment in pursuing professional obligations)

Right to refuse to engage in unethical activity

Right to express one's professional judgment, including the right to dissent

Right to warn the public of dangers

Right to fair recognition and remuneration for professional services

III Employee rights—rights which apply or refer to the status of employees

A Contractual—arising solely out of an employee contract

Example: Right to receive a salary of a certain amount

B Noncontractual—existing even if not formally recognized in a contract or company policy

Examples:

Right to choose outside activities

Right to privacy and employer confidentiality

Right to due process from employer

Right to nondiscrimination and absence of sexual harassment at the workplace

Professional and employee rights can be justified by reference to ethical theories. For example, a rights-based theory would derive the right of professional conscience from a fundamental human right to pursue legitimate interests, where such interests include moral obligations. A duty-based theory might appeal to the fundamental human duty employers have not to harm others (e.g., the public) by handicapping engineers seeking to meet their professional obligations. A goal-based utilitarian theory would argue that the greatest good is promoted by allowing engineers to pursue their obligations. In general, the importance of professional *duties* means that the importance of the *right* to meet those duties must be recognized.

Whistleblowing is most often intentionally conveying new information outside approved organizational channels or against a supervisor's orders with the aim of drawing attention to a significant moral problem. In *external whistleblowing* the information is passed outside the organization, while in *internal whistleblowing* it is conveyed to someone within the organization. In *open* whistleblowing the whistleblowers identify themselves, while in *anonymous* whistleblowing they do not.

Whistleblowing concerning safety is morally *permissible* in those situations where the problem involved is serious and the whistleblowers have first made reasonable attempts to warn others about it through regular organizational channels. It is (prima facie) morally *obligatory* where, in addition, those blowing the whistle are certain they could convince reasonable observers that their views are right and company policy wrong and where there is strong evidence that going public with their information will lead to positive remedies. While in the past the fate of whistleblowers has usually been unhappy, current laws, government policy, and responsible management are moving in the direction of an increased recognition of a limited whistleblowing right.

Both employee rights and professional rights must be subordinate in some respects to the rights of employers to promote company interests. For example, the right to privacy is limited by the need employers have to acquire relevant information about employee skills. But not just any company interest can override employee rights. This is especially clear in regard to the right not to be discriminated against because of sex, race, age, or religion.

Contemporary disagreements over how to deal with discrimination center on the issue of reverse preferential treatment. *Weak preferential treatment* involves giving an advantage to members of traditionally discriminated-against groups over equally qualified applicants who are members of other groups. *Strong preferential treatment* involves giving preference to minority applicants or women over better-qualified applicants from other groups. Arguments for and against such treatment focus on the right to equal employment opportunity, the right to receive and duty to give compensation for past wrongs, and the best way to achieve social integration.

PART IV

CAREER CHOICE AND FURTHER ISSUES

The duty of choosing a profession has been well called . . . the most important of all duties.

Hastings Rashdall

You have got to know what it is you want, or someone is going to sell you a bill of goods somewhere along the line that can do irreparable damage to your self-esteem, your sense of worth, and your stewardship of the talents that God gave you.

Richard Nelson Bolles, from
What Color is Your Parachute?

CHAPTER 7

MORALITY AND CAREER CHOICE

On February 2, 1976, three engineers made news headlines when they resigned from General Electric (G.E.) to protest the nuclear energy industry in which they had worked for years. Each of the engineers was married and had three children. Each left a comfortable job to work full time and without pay for a nuclear protest group. Greg Minor had worked 16 years for General Electric and was manager of advanced control and instrumentation for the nuclear energy division of G.E.'s San Jose plant. Dale G. Bridenbaugh had worked 23 years with General Electric and was manager of performance evaluation and improvement. And Richard Hubbard, at age 38, was manager of quality assurance in the control and instrumentation department.

The three gave as the reason for their resignations the extreme danger they believed to exist in the way nuclear energy was being developed. In his resignation letter Greg Minor expressed their shared sentiments:

> My reason for leaving is a deep conviction that nuclear reactors and nuclear weapons now present a serious danger to the future of all life on this planet. I am convinced that the reactors, the nuclear fuel cycle and waste storage systems are not safe (quoted in Barnett, 34).

Beyond the element of public protest involved, the episode illustrates a dimension of ethics in engineering with which we have not yet dealt. Moral considerations often do and should play a central role in decisions about career choice and career changes. In some ways career decisions are among the most morally important decisions a person makes. At the same time they are very personal decisions. Just as some engineers have decided on moral grounds that they ought not to participate in nuclear energy development, many others have seen supporting it as a moral imperative, considering it as the best solution to the energy crisis and the best way to remove dependence on foreign oil—matters which have wide repercussions for the well-being of our society.

In this chapter we begin by exploring some general connections between morality and career decisions. Following this we survey the moral and social issues confronted by those engaged in engineering within four contexts. Three of these contexts are but selections illustrating the impact of technology on the public: computer technology, industries that pollute the environment, and weapons development. In regard to the fourth context— consulting engineering—we will discuss the work-related moral issues that can arise for self-employed engineers functioning as entrepreneurs within our existing economic system and business milieu.

ETHICS AND VOCATION

At first sight it may seem odd to think of career decisions as having any special connection with morality. In choosing careers we consider prospects of job offers, advancement, security, salary, prestige, life style, personal challenge, and personal satisfaction. These factors apparently have more to do with prudence in seeking personal happiness than with morality.

Perhaps, as Kant thought, there is a kind of indirect moral duty to achieve our own happiness (Kant, 18). If we are depressed and unhappy, we may be unwilling to pursue our moral duties. Or as Erich Fromm, Abraham Maslow, and other recent humanistic psychologists have said, people are unable to be genuinely concerned about others if they are unable to care about themselves. But this line of thought would establish only a very tenuous connection between careers and morality.

Before identifying more direct connections, however, let us elaborate on the aspects of engineering mentioned above which relate to seeking personal happiness.

Existential Pleasures of Engineering

Happiness can be viewed, following Aristotle, as self-realization, rather than mere contentment (Aristotle, 1104). Self-realization, in turn, comes through the exercise of one's highest talents, interests, skills, and virtues. Within limits, the greater the complexity and challenge to one's talents, the greater one's happiness tends to be. Now engineering is a complex and challenging discipline. Moreover, the undergraduate curriculum for engineering is acknowledged to be more rigorous and difficult than the majority of academic disciplines. Combining these generalizations, we might guess that students are attracted to engineering at least in part because of the challenge it offers to intelligent people.

Do empirical studies back up this somewhat flattering portrayal? To a significant extent, yes. Typical students are motivated to enter engineering primarily by a desire for interesting and challenging work. They have an "activist orientation" in the sense of wanting to create concrete objects and systems—to make them and to make them work. They are more intelligent than average college students, although they tend to have a low tolerance for ambiguities and uncertainties which cannot be measured and translated into figures (Perrucci and Gerstl, *Profession Without Community: Engineers in American Society,* 27–52).

What is it that is so appealing and challenging in making technological products? Perhaps no one has so elegantly conveyed the excitement of engineering as Samuel Florman in his book, *The Existential Pleasures of Engineering.* By "existential pleasures" Florman means deep-rooted and elemental satisfactions. He portrays engineering as essentially an attempt to obtain and apply an understanding of the universe so as to fulfill human needs and desires. This attempt calls forth some of the finest aspirations and deepest impulses of human beings.

The first existential pleasure Florman distinguishes resides in the act of personally changing the world. By nature, humans are compelled to improve the world. There is no end to the possibilities for achieving improvements, and the allure of "endless vistas bewitches the engineer of every era" (Florman, 1976, 120–121).

Changing the world brings with it the second type of existential pleasure: the joy of creative effort. This includes planning, designing, testing, producing, selling, constructing and maintaining. In contrast with the scientist's main interest in discovering new knowledge, the engineer's greatest enjoyment derives from creatively solving practical problems (Florman, 1976, 143).

Yet the engineer shares, as a third type of pleasure, the scientist's joy in understanding the laws and riddles of the universe. Both may experience "quasi-mystical moments of peace and wonder" (Florman, 1976, 141).

A fourth pleasure relates specifically to size in the world. The magnitude of natural phenomena—oceans, rivers, mountains, and prairies—both intimidates and inspires. In response, engineers conceive immense ships, bridges, tunnels, and other "mammoth undertakings [which] appeal to a human passion that appears to be inextinguishable" (Florman, 1976, 122).

A fifth pleasure relates to regularly being in the presence of machines. A mechanical environment can generate a comforting and absorbing sense of a manageable, controlled, and ordered world. "For a period of time, personal concerns, particularly petty concerns, are forgotten, as the mind becomes enchanted with the patterns of an orderly and circumscribed scene" (Florman, 1976, 137).

Concern for Humanity

Florman concludes his inventory of the existential pleasures of engineering with what he says is the primary and most important one of all: "a strong sense of *helping,* of directing efforts toward easing the lot of one's fellows" (Florman, 1976, 145). "The main existential pleasure of the engineer," he writes, "will always be to contribute to the well-being of his fellow man" (Florman, 1976, 147).

This paramount source of existential pleasure and motivation suggests one straightforward connection between morality and career decisions in engineering. Virtually every traditional characterization of the nature and goals of engineering has emphasized the contribution it makes to improving human life. This is not mere propaganda unrelated to what really concerns engineers on the job. The same theme is sounded in the personal testimony of many engineers as they reflect on what their careers have meant to them (Florman, 1976, 94–95).

This needs to be borne in mind when interpreting empirical studies of students' main motives for entering engineering. The technical challenge it offers is always seen by them against the background of participating in a socially useful and important enterprise. Engineering would not attract students if it were viewed by them as generally directed toward immoral ends.

Our interest here is in the possibility of this implicit background of moral concern becoming more explicit in thinking about career decisions. One might well ask what moral aims one's career has or might have. From such a perspective, ethics does not need to be viewed negatively as a burden or a constraint on one's career. Rather it offers positive ideals for expressing natural moral interests in concrete ways within one's professional life. We are led to consider what has been called the ethics, or moral philosophy, of vocation.

Philosophy of Vocation

William Frankena, a leading contemporary ethicist, has attempted to state the fundamental tenets of a moral philosophy of vocation. Four of his suggestive ideas may be summarized as follows (Frankena, 1976, 393–408).

First, according to Frankena, each person has the important duty of selecting a vocation. Most vocations are jobs which enable people to earn a living, whether or not a given individual needs the income from his or her job. But they are also, and more essentially, types of work which enable people to find senses of identity, personal worth, and meaning, and which promote feelings of contributing to the good of others. Thus the work of raising children qualifies as a vocation, even though in fact no income is usually obtained from it. Selecting a vocation is a duty, according to Frankena, both because of the moral importance of seeking self-fulfillment and because of the good that can be achieved for others through pursuing a vocation.

Second, vocations carry with them a set of specific duties. Fulfilling those is the main way in which a person fulfills the general duty of doing good for others. Utilitarians might say that the amount and kind of good one has a duty to produce is largely determined by the nature of one's vocation. This could be called "vocation-utilitarianism," in which good consequences are weighed largely in respect to the role of a vocation rather than in respect to individual acts (act-utilitarianism) or general rules (rule-utilitarianism).

Frankena is not a utilitarian, but as a deontologist (i.e., duty-based ethicist) he admits that one fundamental duty is the principle of benefi-cence: to create good and prevent harm. The amount of good people are obligated to create is partly determined by the nature of their vocations. This is because concrete moral obligations are specified in part by a direct appeal to professional duties, as well as by an appeal to foundational ethical theories. Presumably professional duties are themselves grounded in the more general ethical principles, as we have suggested throughout this book.

Third, while there is a duty to have a vocation, there is normally no specific vocation one is obligated to pursue. Rather, people should be free to select their own careers as an aspect of their autonomous self-expression. This is because both personal self-realization (what Frankena calls "the good life") and the morally good life are based on autonomous decision making.

Fourth, and last, there are objective considerations which ought to be weighed in making career choices. Most important are the prospects of happiness and self-fulfillment. One must ask which available vocation, or type of vocation, it would be rational to choose in order to best express one's main talents and basic interests. Also very important, however, are the moral values to which one is committed. For it is largely in and through one's vocation that those values will be expressed and realized.

These four tenets can be developed with respect to the choice of engineering as a vocation. No one is obligated to become an engineer, for that should be a matter of free personal decision. Yet the decision to enter engineering should be made on the basis of a searching assessment of one's talents and interests, and also on the basis of relating engineering to one's basic moral values.

Engineering involves a particular set of professional duties specified in terms of practical ways to prevent harm and promote good. A reasonable decision to enter engineering entails a prior weighing of those duties against one's general values and ideals. Does an engineering career promise an avenue for self-realization and is it compatible with one's moral concern?

Frankena's line of thought can also be extended to decisions about entering branches of engineering and areas of engineering work. To use an example already mentioned, people contemplating a career having to do with the development of nuclear energy ought to investigate the types of moral issues and dilemmas they might have to confront. A serious investigation may involve course work related to the social ramifications of nuclear engineering.

This last point relates to the implications of a philosophy of vocation for education. Frankena says his views require that education be both liberal and vocational. It should be vocational in the sense of preparing people for the vocations he thinks they are obligated to pursue. But it should be a liberal arts education as well, in order to enable them to choose their vocations autonomously and with wisdom. Education should foster both self-knowledge and serious reflection upon the nature and function of values.

Work Ethics

Frankena's views provide the theoretical basis for what is commonly called a *work ethic*. This idea calls for comment. There is, in fact, no one work ethic. There are many different ones which different people embrace, and no doubt there are several morally legitimate and some unjustified perspectives on work (Cherrington, 19–30).

One version was identified by the sociologist Max Weber in *The Protestant Ethic and The Spirit of Capitalism*. The Protestant work ethic, according to Weber, provides the psychological explanation for the rise of modern capitalism. In brief, it was the idea that financial success is a sign that predestination has ordained one as favored by God. This was thought to imply that making maximal profits is a duty mandated by God. Profit becomes an end in itself rather than a means to other ends. It is to be sought rationally, diligently, and without compromise with other values.

Carried to an extreme, this idea can lead to the life and ideology of the workaholic. Whether out of religious or secular concerns, workaholics are compulsive overworkers. Their entire energies are devoted to work, to the point of unhealthy addiction. As with other addicts, workaholics are said to be pushed by uncontrolled and unconscious urges: anxiety, guilt, insecurity, or a sense of inferiority. As a result their behavior is irrational. They neglect their families, have few personal relationships not tied to their work, and seem unable to fill leisure time with meaningful activity. Death tends to come early, often from heart attack (Cherrington, 253–274).

Focusing on this irrational extreme should not lead to a rejection of the possibility of a reasonable work ethic. There is a difference between a compulsive workaholic and a healthy hard worker who also has a wide range of additional interests. Major professions, like engineering, make substantial demands upon those engaged in them which require the virtues of discipline, initiative, and keeping up to date in regard to knowledge and skills. To this extent professional ethics merges with a work ethic which emphasizes attitudes and virtues needed for fulfilling professional obligations.

Beyond this point, there is a range of general attitudes which may enter into an individual's personal work ethic. Several of those attitudes might be summarized as follows:

1 Work is intrinsically valuable to the extent that it is enjoyable or meaningful in allowing personal expression and self-fulfillment. Meaningful work is worth doing for its own sake and for the sense of personal identity and self-esteem it brings.

2 Work is the major instrumental good in life. It is the central means for providing the income needed to avoid economic dependence on others, for obtaining desired goods and services, and for achieving status and recognition from others.

3 Work is a necessary evil. It is the sort of thing one must do in order to avoid worse evils, such as dependency and poverty. But it is mind-numbing, degrading, and a major source of anxiety and unhappiness.

As we noted earlier, most people enter engineering for its inherent challenge. This presupposes that they seek a vocation warranting attitudes in the first category, although the second type of attitude is also important. They hope to find in engineering a vocation which evokes the pride and self-esteem of the craftsperson. Meaningful tasks are to be accomplished because of their intrinsic importance, not because a supervisor is watching.

Models of Professional Roles

Models and metaphors often serve to organize thinking and crystallize attitudes. It was with this in mind that we earlier suggested the model of engineering as social experimentation and the engineer as an experimenter. Many other models have been advocated in connection with the social and moral roles of engineering and engineers. In concluding this section, we list some of those models which at times have seemed attractive to some engineers. We leave as a study question the assessment of the models in connection with thinking about career goals. Those who wish to examine the influence of several of them in the first half of this century are referred to Edwin Layton's masterful study, *The Revolt of the Engineers.*

1 *Savior.* Plato believed a philosopher-king was needed in order to create the ideal society. Others have believed that engineers hold the key to creating a utopian society. In part this is to be achieved through technological developments which lead to material prosperity. In part it is to arise through creating a technocracy in which engineering ways of thinking are applied to large-scale social planning. The representative engineer is a savior who will redeem society from poverty, inefficiency, waste, and the drudgery of manual labor.

2 *Guardian.* Perhaps engineers cannot usher in utopia. Yet it is they who know best the directions in which, and pace at which, technology should develop. Accordingly, they should be given positions of high authority based on their expertise in determining what is in the best interests of society (Veblen, 52–82).

3 *Bureaucratic servant.* Within the corporate setting in which engineers work, management should make the decisions about the directions of technological development. The proper role of the engineer is to be the servant or handmaiden who receives and translates the directives of management into concrete achievements. The engineer is the loyal organization man or woman whose special skills reside in solving problems assigned by management, within the constraints set by management.

4 *Social servant.* The role of engineers lies exclusively in obedient service to others, but their true master is society. Society expresses its interests either directly through purchasing patterns, or indirectly through government representatives and consumer groups. Engineers in cooperation with management have the task of receiving society's directives and satisfying society's desires.

5 *Social enabler and catalyst.* This is a variation on model 4. Service to society includes, but is not exhausted by, carrying out social directives. Ultimate power and authority lie with management, but nevertheless the engineer plays a vital and active role beyond mere order-following. Sometimes engineers are needed to help management and society understand their own needs and to make informed decisions about desirable ends and means of technological development (Fruchtbaum, 258).

6 *Game player.* Engineers are neither servants nor masters of anyone. Instead they play by the economic game rules that happen to be in effect at a given time. Their aim, like those of managers, is to play successfully within organizations, enjoying both the pleasures of technological work and the satisfaction of winning and moving ahead in a competitive world (Maccoby).

Study Questions

1 Assess the positive and negative implications of each of the following models for engineers: savior, guardian, bureaucratic servant, social servant, social enabler and catalyst, and game player. Which would you say best captures the type of engineer you would like to be or would like engineers to be?
2 Do you agree with Frankena that there is a duty to have a vocation, in his sense of the term "vocation?" Respond to the criticism that his view is excessively moralistic.
3 Formulate a work ethic which expresses your own attitudes about work.

COMPUTER ETHICS

As the first example of professional activity and its ethical implications we shall now examine the growing use of computers, a development which has sufficiently unique problems to allow us to address them in a separate section on computer ethics.

Computers are becoming the technological backbone of our society. Their degree of sophistication, range of applications, and sheer numbers are rapidly increasing. Yet electronic computers are still only a few decades old, and it is difficult to foresee all the moral problems which will eventually surround them. The present state of computers is sometimes compared to that of the automobile in the early part of this century. At that time the impact of cars on work and leisure patterns, pollution, energy consumption, and sexual mores was largely unimagined. If anything, it is more difficult to envisage the eventual impact of computers since they are not limited to any one primary area of use comparable to a car's function in transportation.

It is already clear, however, that computers will generate or contribute to a variety of moral problems. Because of this, it is appropriate to speak of *computer ethics* as that branch of engineering ethics dealing with moral issues in computer technology. Some of these issues concern abuse of existing installations, as when computers are used in criminal activities and invasions of privacy. Others relate to assessments of the desirable directions of new computer applications. All such issues are of interest both to engineers, who will work with or have close contact with computer technology, and to the public, which is increasingly confronted directly or indirectly with a computerized world. Some of the following problems and examples are similar to ones we have discussed earlier, but here the computer connection gives them a special flavor.

Computer Abuse

The term "computer abuse" does not refer merely to the damaging of computers, although it includes that kind of problem. As we will define it, computer abuse is any unethical or illegal conduct in which computers play a central role (whether as instruments or objects), as opposed to their playing an incidental or peripheral role. The distinction between central and incidental roles is somewhat rough, but a few examples will help to clarify it.

Computers are only incidentally involved when extortion is attempted via a phone which is part of a computerized telephone system (Kling, 1980, 408). By contrast, computers are centrally involved when an extortionist disguises his voice by means of a computer as he talks into a phone. And computers are even more centrally involved when an unauthorized person uses a telephone computer system to obtain private phone numbers, or when someone maliciously alters or scrambles the programming of a telephone computer, or when a private or governmental hit-squad uses computers to identify fingerprints of its intended victims.

Computer Theft and Fraud

Some of the most commonly discussed cases of computer abuse are instances of outright theft and fraud. There are many forms of computer theft and fraud. For example: (1) stealing or cheating by employees at work; (2) stealing by nonemployees or former employees; (3) stealing from or cheating clients and consumers; (4) violating contracts for computer sales or service; (5) conspiring to use computer networks to engage in widespread fraud; (6) advertising "next-generation" computers in a misleading manner; (7) extortion practiced by the sole programmer or operator of a computer installation who has a unique knowledge of how it works or is programmed.

Public interest has often been drawn to the glamorous capers of computer criminals (Whiteside). Enormous sums of money have been involved. The average amount stolen in conventional embezzlement is $19,000. The amount for an average computer-related embezzlement is $430,000, and many millions are often involved. Yet the giant thefts uncovered are believed to be only a small fraction of computer theft—less than 1 percent (Logsdon, 163–164).

Crime by computer has proven to be unusually inviting to many. Computer crooks tend to be intelligent and to view their exploits as intellectual challenges. In addition, the computer terminal is both physically and psychologically far removed from face-to-face contact with the victims of the crimes perpetrated. Unlike violent criminals, computer criminals find it easy to deceive themselves into thinking they are not really hurting anyone, especially if they see their actions as nothing more than pranks. In addition there are often inadequate safeguards against computer crime. The technology for preventing and catching criminals has lagged behind implementation of new computer applications. Computers reduce paper-work, but this has the drawback of removing the normal trail of written evidence involved in conventional white collar crime (forgeries, receipts, etc.). Finally, the penalties for computer crime, as for white collar crime in general, are mild compared to more conventional crimes.

Computer crime raises obvious moral concerns related to basic issues of honesty, integrity, and trust. It also forces a rethinking of public attitudes about crime and its punishment. Is it fair that the penalty for breaking into a gas station and stealing $100 should be the same as for embezzling $100,000 from a bank? How should society weigh crimes of minor violence against nonviolent crimes involving huge sums of money?

The potential for computer crime should enter significantly into the thinking of engineers who design computers. In fact, protection against criminal abuse has become a major constraint for effective and successful design of many computer systems and programs. Engineers must envisage not only the intended context in which the computer will be used, but both likely and possible abuses.

For some time secret computer passwords have been used as a security feature. More recently, and still of limited effectiveness, is data encryption. This technique is widely employed to prevent theft from funds transfer systems. In data encryption, messages are scrambled before transmission and unscrambled after reception according to secret codes. Such devices, of course, require special precautions in maintaining confidentiality and security, and engineers have a major role to play in making recommenda-tions in these areas.

Privacy

Storage, retrieval, and transmission of information using computers as data processors has revolutionized efficient communication. Yet this very benefit poses serious moral threats to the right to privacy. By making more data available to more people with more ease, computers make privacy more difficult to protect. Here we will discuss privacy and confidentiality in regard to individuals, but the issues are similar in relation to corporations.

In the Chap. 6 discussion of employee rights, we defined the right to privacy as the right to control access to and use of information about oneself. We indicated that the right was vitally important for a variety of reasons, and could be grounded within each of the major types of ethical theories. Here we will take note of some of the threats to that right from computerized data centers.

Imagine that you are arrested for a serious crime you did not commit—for example, murder or grand theft. Records of the arrest, any subsequent criminal charges, and information about you gathered for the trial proceedings might be placed on computer tapes easily accessible to any law enforcement officer in the country. Prospective employers doing security checks could gain access to the information. The record clearly indicates that you were found innocent legally. Nevertheless that computerized record could constitute a standing bias against you for the rest of your life, at least in the eyes of many people with access to it.

The same bias could exist if you had actually committed some much less serious crime—say a misdemeanor. If you were arrested when you were 15 for drinking alcohol or swearing at an officer, for example, the record could stay with you. Or imagine that medical data about your visits to a psychiatrist during a period of depression could be accessed through a data bank. Or that erroneous data about a loan default were placed in a national credit data bank to which you had limited access. Or merely suppose that your tastes in magazine subscriptions were known easily to any employer or ad agency in the country.

The potential abuses of information about us are unlimited and become more likely with the proliferation of access to that information. For this reason a series of new laws has been enacted (Rule, McAdam, Stearns, and Uglow). For example, the 1970 Fair Credit Reporting Act restricted access to credit files. Information can be obtained only by consumer consent or a court order, or for a limited range of valid credit checks needed in business, employment, and insurance transactions or investigations. The act also gave consumers the right to examine and challenge information about themselves contained in computerized files.

The Privacy Act of 1974 extended this right of inspection and error correction to federal government files. It also prohibited the information contained in government files from being used for purposes beyond those for which it was originally gathered unless such use was explicitly agreed to by the person whose file it is. Numerous other laws have been passed and are being considered to extend protection of individual privacy within private business and industry.

Such laws are expensive to implement, sometimes costing tens and hundreds of millions of dollars. They also lessen economic efficiency. In special circumstances they can have harmful effects on the public. There is little question, for example, that it would save lives if medical researchers had much freer access to confidential medical records. And it would be much more convenient to have one centralized National Data Center. This idea was proposed in the mid-1960s and is still alive in the minds of many. But privacy within a computerized world can apparently be protected only by making it inconvenient and expensive for others to gather data bank information on us.

Further Social Issues

Crime and privacy have been the focus of most discussions about computer ethics. More attention, however, is now being given to a variety of other social and professional issues. Let us survey some of these, beginning with social issues.

1 Selective disclosure of their views by politicians has always occurred. In a speech to a conservative group a candidate for senator will tend to say very different things from what he or she tells a liberal group. Different topics may be discussed, different emphases given, and inconsistent remarks made. Computers make it possible to turn this political maneuver into a science. Dozens of types of letters can be worded to appeal to different kinds of groups. Each type reveals only a fraction of the candidate's views—just that portion which will be most attractive to a selected sociopolitical group.

The information about these groups of people is obtained by computer from public records. The characterizations of the groups' attitudes and norms are computer-generated. The letters sent to them are personalized by computer. And the mailing process is also computerized. With electronic accuracy and efficiency, politicians are enabled to have many different faces when viewed by different groups.

Several moral issues are raised by this possible application of technology: (1) Does such selective disclosure constitute deception? (2) Does filtering the truth about a politician's views undermine the autonomy of voters in making decisions? (3) Since use of computers is expensive, is it fair that the rich have more extensive access to this technology, or should equal-time laws for television be extended to computers? (Maner, 2–8).

2 Many new occupations have been created by computers: data processing, programming, information science, key punch operating, computer systems analysis, computer sales positions, etc. Should the importance of responsible behavior by these groups lead the public to seek their full professionalization and require licensing, registration, or continuing education? What should the public demand by way of accountability from professional organizations representing these groups?

3 Most existing professional groups are finding their work altered by computers. This will have subtle effects on areas of personal accountability. For example, it was once clear that doctors were fully responsible for the diagnoses they made: Errors in judgment were *their* errors. Increasingly, however, diagnoses and other medical decisions are and will come to be made with varying degrees of computer assistance. Health data on patients may be collected with computerized instruments and then entered into a computer which is programmed to diagnose illness on the basis of the medical symptoms entered. Under such conditions, when would a doctor be responsible for a mistaken diagnosis due to computer error? When would a doctor be blameworthy for overreliance on a mistaken computer diagnosis?

4 Computers have led and will continue to lead to elimination of jobs. What employer attitudes are desirable in dealing with this situation? No employer, of course, can afford to pay people for doing no work. Yet especially within large corporations it is often possible to readjust work assignments and work loads and to wait for people to retire or to change jobs voluntarily before laying off employees. Such benign employment practices have often been embraced out of prudential motivation to prevent a public and employee backlash against introducing computer technologies which eliminate jobs (Kling, 1979, 10). But there are also moral considerations relating to human costs which must be weighed.

5 There are also questions concerning public accountability of businesses using computer-based services. It can be made very difficult or relatively simple for a consumer to notice and correct computer errors or computer-printed errors. For example, a grocery-store receipt can itemize items by obscure symbols or simple words understandable to a customer. Here again moral reasons reinforce long-term good business sense in favoring policies beneficial to consumer needs and interests.

Professional Issues

Many of the issues in engineering ethics which we have dealt with earlier arise again within the context of computer work. New variations or new difficulties may be involved, often owing to the high degree of job complexity and required technical proficiency introduced by computers. Such was the case, for example, in the whistleblowing case of Virginia Edgerton given as a study question in Chap. 6. A number of interesting case studies have been collected by Donn Parker in his excellent book, *Ethical Conflicts in Computer Science and Technology*. The following examples are based on some of Parker's cases.

Safety Dependence on computers has intensified the division of labor within engineering. For example, civil engineers designing a flood control system to rely on information and programs obtained from systems analysts and implemented by computer programmers. Suppose the systems analysts refuse to assume any moral or legal responsibility for the safety of the people affected by the flood control plans, arguing that they are merely providing tools whose use is entirely up to the engineers. Should the civil engineers be held accountable for any harm caused by poor computer programs? Presumably their accountability does extend to errors resulting from their own inadequate specifications which they supply to the computer experts. Yet should not the engineers also be expected to contract with computer specialists who agree to be partially accountable for the end-use effects of their program? (Parker, 34–38)

Ownership Rights Consider an engineer who develops a program used as a tool in developing other programs assigned to her. Subsequently she changes jobs and takes the only copy of the first program with her for use on her new job. Suppose first that the program was developed on company time under the first employer's explicit directives. Taking it to a new job without the original employer's consent would be a violation of that employer's right to the product (and possibly a breach of confidentiality). As a variant situation, however, suppose the program was not written under direct assignment from the first employer, but was undertaken by the engineer at her own discretion to help her on her regular work assignments. Suppose also that to a large extent the program was developed on her own time on weekends, although she did use the employer's facilities and computer services. Did the employer own or partially own the program? Was she required to obtain the employer's permission before using it on the new job? (Parker, 72–74)

Whistleblowing An engineer working as a computer programmer played a minor role in developing a computer system for a state department of health. The system involved storing medical information on individuals identified by name. Through no fault of the engineer, few controls had been placed on the system to limit easy access to it by unauthorized people. Upon learning of this the engineer first informed his supervisor and then higher management, all of whom refused to do anything about the situation because of the anticipated expense required to correct it. In violation of the rules for using the system, the programmer very easily obtained a copy of his own medical records. He then sent them to a state legislator as evidence for his claims that the right of citizens to confidentiality regarding such information was threatened by the system. Was his behavior improper? Was his subsequent firing justified? (Parker, 90–93)

Employer Authority and Professional Rights A project leader working for a large retail business was assigned the task of developing a customer billing and credit system. The budget assigned for the project appeared at first to be adequate. Yet by the time the system was half completed it was clear the funds were not nearly enough. The project leader asked for more money, but the request was denied. He fully informed management of the serious problems which were likely to occur if he had to stay within the original budget. He would be forced to omit several important program functions relating both to convenience and to safety: for example, efficient detection and correction mechanisms for errors, automatic handling and reporting of special customer exceptions, and audit controls. Management insisted that these functions could be added after the more minimal system was produced and installed in stores. Working under direct orders, the project leader completed the minimal system, only to find his worst fears realized after it was installed. Numerous customers were given incorrect billings or ones they could not understand. It was easy for retail salespersons to take advantage of the system to steal from the company, and several did so. Within a year the company's profits and business were beginning to drop. This led to middle-level management changes, and the project leader found himself blamed for designing an inadequate system.

Did the project leader have an obligation either to clients or to the company to act differently than he did? Did he have a moral right to take further steps in support of his original request or later to protect himself from managerial sanctions? (Parker, 109–111)

Informed Consent A team of engineers and biomedical computer scientists develop a system for identifying people from a distance of up to 200 meters. A short tube attached to a sophisticated receiver and computer, and aimed at a person's head, reads the individual's unique pattern of brain waves when standard words are spoken. The team patents the invention and forms a company to manufacture and sell it. The device is an immediate success within the banking industry. It is used to secretly verify the identification of customers at tellers' windows. The scientists and engineers, however, disavow any responsibility for such uses of the device without customer notification or consent. They contend that the companies which buy the product are responsible for its use. They also refuse to be involved in notifying public representatives about the product's availability and the way it is being used.

Does employing the device without customer awareness violate the rights to privacy or informed consent? Do the engineers and scientists perhaps have a moral obligation to market the product with suggested guidelines for its ethical use? Should they be involved in public discussions about permissible ways of using it? (Parker, 126–128)

Study Questions

1 Present and defend your answers to the questions raised about the cases based on Parker's examples.
2 Look up the following terms in a legal dictionary: fraud, theft, extortion, sabotage, vandalism, burglary. Then imagine and briefly describe an example of a computer crime falling under the definition of each.

3 Write a short research paper exploring the threats to privacy posed by data banks. In your essay comment on some specific advantages and disadvantages of having one centralized national data bank which pools all available government information on citizens.

4 The U. S. Office of Technology Assessment has reported that new rules may soon be needed for computer services concerning the prediction of criminal and financially negligent behavior. People are likely in the future to be denied credit, employment, and insurance because they fall into certain categories based on statistical correlations developed by computers.

Comment on who you think should establish these new rules, and give any examples of rules you would favor. For example, do you think it would be fair to allow race or sex to be used as a variable in this context?

5 Tracy Kidder's Pulitzer Prize winning book, *The Soul of a New Machine,* recounts the human drama behind the development of a minicomputer by Data General Corporation (Kidder). Read the book, and as you do so, identify any moral issues which may arise in the course of creating a new computer system and bringing it to market.

THE ENVIRONMENT

Our next set of examples is taken from technological endeavors which affect the environment. Consider the following episode:

> Roads began cracking like china. Railroad tracks buckled. Sewer lines burst. Oil drillers saw their pipes twist and bend like spaghetti. Pacific tides washed over harbor bulkheads. . . . (Marx, 140)

An earthquake? Not quite. Just a gradual but steady land subsidence, inadvertently caused by human beings, played back at rapid speed. Wesley Marx continues as follows in his book *Acts of God, Acts of Man* with a description of the geologic subversion of Long Beach during 1941 to 1945:

> The Long Beach naval shipyard wondered if it could stay intact long enough to help end World War II. Some areas sank as much as twenty-nine feet, or enough to reach the groundwater table. . . . Gravity no longer provided drainage free of charge. . . . When would the sinking stop? Some industrialists didn't care to find out. They were prepared to climb out of their inverted landscape and never return (Marx, 140).

When it was discovered that extensive pumping of oil fields in the vicinity was draining the underground reservoirs and reducing underground pressure, water was reinjected into those reservoirs and the lowering of the land surface was greatly reduced. Water pumping can cause similar subsidence, as many cities and areas have experienced, from Tokyo, Osaka, and Nagoya in Japan to the 3000-square-mile area between Houston and Galveston, Texas. Houses will then lean and lowered land will experience severe flooding during typhoons and hurricanes.

Water is injected underground not only to prevent subsidence, but also to force oil out of aging wells. In 1963, such high-pressure injection in oil fields near Baldwin Dam in Los Angeles caused the bottom of the reservoir to crack open along an old fault line. The crack reached all the way to the foundation of the dam. The 12-year-old dam had been designed to withstand earthquakes but not such surprises. The released waters killed five people and did $14 million worth of damage to property (Marx, 141; Sowers and Sowers, 535).

Engineers have played an important role in creating a technology which lessens our physical burdens, protects us from hostile environments, and expands our opportunities for enjoying life. But now we are becoming increasingly aware of technology's "side effects," many of which are not as visible or obvious as shifting ground.

Some observers warn us that we are misusing our scarce resources, fouling our environment, and in general practicing growth in consumption and population which will eventually make "spaceship earth" too small for us. Others think the situation may not be so bleak. Nature itself, they point out, is continually changing the face of the earth and human activity creates at best a small perturbation. But it is generally agreed that industrial activity does affect the biosphere: It denudes the land, pollutes the waters and the atmosphere, and threatens fragile species. But how damaging such effects are in the long run, or to what extent we can develop countermeasures in time, is difficult to assess. As Herman Kahn, William Brown, and Leon Martel point out:

> It may yet turn out that future man will marvel at the paradoxical combination of hubris and modesty of 20th century man, who, at the same time, so exaggerated his ability to do damage and so underestimated his own ability to adapt to or solve such problems. Or it could be that future man, if he exists, will wonder at the recklessness and callousness of 20th century scientists and governments (Kahn, Brown, and Martel, 180).

While we do not share the pessimists' views of inevitable doom, we also do not subscribe to the optimists' views that new technologies will invariably spring up to deliver us from the traps set by the older technologies. We do recognize that we cannot turn back the clock, that we have to live with some unalterable changes and have already learned to do so. Nevertheless, we observe that technology and its practitioners have engaged us in an ongoing global experiment that needs to be monitored and guided lest we permit it to lead us into many unacceptable changes of an irrevocable nature.

Engineers need to be aware of their role as agents of change, as experimenters. There are questions every engineer should ask, whether at the time of choosing a career or when entering a new sphere of activity. To what extent does a particular industry affect the environment? To what extent can those effects be controlled physically or regulated politically? Have all reasonable abatement measures been implemented? Can I be effective as an engineer in helping to assure a decent environment for the generations that will follow? What are my responsibilities in all this?

Examples

Of the myriad environmental issues of concern to engineers we will consider only three. The first has to do with the use of toxic substances in food processing, the second with release of harmful substances into air and water, and the third with upsetting a delicate ecosystem. Following these illustrative cases, we will touch upon the economic and political means of controlling the dangers.

PCBs and Kanemi's Rice-Oil—Japan, 1968   A decade of rapid industrial growth in Japan had taken its toll of the environment. City dwellers fell ill from air pollution. Some rivers were covered with dead fish floating on the surface. The dreadful Minamata disease, traceable to mercury pollution of a nearby bay by the Chisso Company and a foodchain involving shellfish taken from that bay, was continuing to fell its victims (46 deaths and 75 disabilities by 1970). In May the itai-itai ("ouch-ouch") disease, which is painful and causes bone crumbling, was the first illness to be designated as pollution-generated by the Japanese government. Observed on and off since 1920 at various sites, the latest outbreak was placed at the doors of a Mitsui Mining and Smelting facility which let cadmium escape into a river, whence the pollutant found its way into the foodchain via rice paddies and rice. The thalidomide drug tragedy was still on everyone's mind. The severely malformed children of women who had taken that drug during their pregnancies were still not cared for by either the Dainippon Pharmaceutical Company or the government, and schools refused to accept them (Ui; Iijima).

Then, in the summer of 1968, a disease of unknown origin made its appearance in southern Japan. Victims suffered from disfiguring skin acne and discoloration, from fatigue, numbness, respiratory distress, vomiting, and loss of hair. Eventually 10,000 people were stricken and some died. What was the cause? An investigation of 121 cases was conducted, with 121 healthy individuals matched to the victims by age and sex being used as a control group. All 242 were questioned regarding their diets, personal habits, and places of work. When it was discovered that the only significant difference between the two groups was in the amount of fried foods eaten, the disease was traced to rice oil produced by the Kanemi Company (Grundy, Weisbrod, and Epstein, 111).

It took another 7 months to find the specific agent in the oil. Autopsies performed on victims revealed the presence of PCBs, polychlorinated byphenyls (Iijima, 268). Oil made from rice bran at the Kanemi plant was heated at low pressure to remove objectionable odors. The heating pipes were filled with hot PCB fluid, but the pipes were corroding and tiny pinholes in them allowed PCBs to leak directly into the oil. In fact, Kanemi had been in the habit of replenishing about 27 kg of lost PCBs a month for some time.

There are other, less direct paths in which the toxic PCBs can reach humans. For example, the Kanemi rice oil had also been used as an additive to chicken feed. In early 1968, 1 million chickens were given this feed and every second chicken died. In the United States 140,000 chickens were slaughtered in the state of New York on one occasion when data collected by the Campbell's Soup Company revealed more than the permitted level of PCBs in chickens raised by certain growers. The source of PCBs was found to be plastic bakery wrappers mixed in with ground-up stale bread from bakeries used as feed. On another occasion PCBs leaked from a heating system into fishmeal in a North Carolina pasteurization plant. About 12,000 tons of fishmeal were contaminated and 88,000 chickens, already fed this fishmeal, had to be destroyed when the product was recalled (Grundy, Weisbrod, and Epstein, 112).

PCBs are used not only in heat exchangers. They are a good hydraulic fluid and an excellent dielectric. They also decompose very slowly. These properties made them a good choice for insulating oil in capacitors and transformers. But they are no longer considered suitable for such applications—or in any other where they can find their way into the environment. Their hardiness accounts for the fact that they are found in the oceans in larger quantities than DDT although they enter the oceans initially in much lesser amounts (Murdoch, 286–287). Even a total shutdown of all possible sources of PCB contamination—responsible for about 28,000 tons of PCBs escaping into air, land, and water every year—would not result in an immediate decrease in its presence. Meanwhile, even the present burden of PCBs on the environment is not fully understood (Grundy, Weisbrod, and Epstein, 112–113).

Asbestos in the Air; Asbestos in Drinking Water?   Although the use of PCBs in food and feed processing plants might not have been a problem in the above instances if the heat transfer systems employing the toxic oil had been secure and closed, one should never be as optimistic as to expect a leakproof piping system; and when foodstuffs are involved, one should not even contemplate a substantial use of toxic materials for any reason. There are other processes, however, where even an approximation to a closed system could have great benefits. One such example is ore processing as practiced by the Reserve Mining Company, a subsidiary of Armco and Republic steel companies, when it first started its huge taconite (iron ore) plant at Silver Bay, Minnesota in 1955 (Bartlett; Lawless, 297–307).

Reserve was invited to open operations by the state of Minnesota; it employed over 3000 workers; and it was also making a contribution to the nation by utilizing abundant but low-grade ore to produce much needed iron and steel. Taconite contains only 25 to 30 percent iron oxide. It is crushed and washed in steps to form a powder from which the iron is extracted by electromagnets. Everything at Reserve was on a large scale— size, cost ($350 million), power consumption, and water usage. Rather than utilizing settling basins and recycling the used water as is done in smaller plants, fresh water was drawn from Lake Superior and waste water was discharged into it at the rate of half a million gallons per minute. Despite assurances from the University of Minnesota's experiment station that the fine taconite powder would settle out, it did not. The lake became discolored by the discharge of about 67,000 tons of taconite tailings a day, and fishers complained about the disappearance of fish.

The company was not doing anything very different from what industry had always been doing. In Sweden, paper mills were polluting the Baltic Sea, in the Soviet Union unique Lake Baikal was being similarly disfigured. But the managers of Reserve had not counted on the change in times. In the 1960s concerns about the environment were gradually being taken more seriously. The federal government and eventually the state of Minnesota took the company to court since it would not voluntarily change its practices and retrofit its extraction process to a closed system with no discharge of wastes into the lake. Private citizens had become active and were influential in causing the government agencies to act.

The court proceedings, including appeals, were lengthy (14 years) and replete with contradictory testimony from experts representing the opposing sides. It was not until asbestoslike fibers were found to be entering the air and the lake from the Reserve operations that a more definite link to a real hazard could be established. However, the amounts entering the drinking water supplies of cities ringing Lake Superior, and the levels considered to be harmful, could not be established with any degree of precision. Accordingly, Reserve's taconite beneficiation plant was not closed down but ordered to switch over to a land disposal site as soon as possible.

Too Little Water for the Everglades The great marshes of southern Florida have attracted farmers and real estate developers since the beginning of the century. When drained, they present valuable ground. From 1909 to 1912 a fraudulent land development scheme was attempted in collusion with the then U.S. Secretary of Agriculture. Arthur Morgan blew the whistle on that situation, jeopardizing not only his own position as a supervising drainage engineer with the U.S. Department of Agriculture, but also that of the head of the Office of Drainage Investigation. An attempt to drain the Everglades was made again by a Florida governor from 1926 to 1929. Once more Arthur Morgan, this time in private practice, stepped in to reveal the inadequacy of the plans and thus discourage bond sales.

But schemes affecting the Everglades were not over yet. Beginning in 1949, the U.S. Army Corps of Engineers started diverting excess water from the giant Lake Okeechobee to the Gulf of Mexico to reduce the danger of flooding to nearby sugar plantations. As a result the Everglades, lacking water during the dry season, were drying up. A priceless wildlife refuge was falling prey to humanity's appetite. In addition, the diversion of waters to the Gulf and the ocean also affected human habitations in southern Florida. Cities which once thought they had unlimited supplies of fresh ground water found they were pumping salt water instead as ocean waters seeped in (Morgan, 370–389).

Southern Florida represents a complex environmental unit with a delicate balance. Any intrusion by human engineering must be seen as an experiment which must be conducted with great care. Unfortunately too many public agencies view any change in plans as unacceptable once a course has been charted. As Arthur Morgan points out in his book *Dams and Other Disasters,* the Corps was particularly prone to such an attitude fostered by the crisis-oriented training quite properly offered for other purposes at West Point Military Academy (Morgan, 37).

Guilty until Proven Innocent?

The examples just given barely cover a fraction of the many environmental issues which might arise in engineering practice. But they suffice to raise some key questions of ethical import: *Who* is affecting *whom*—and *where, when,* and *how?*

In the mercury, cadmium, and PCB pollution cases in Japan the question was: Who is releasing toxic substances? In all instances the polluting companies refused to acknowledge their mistakes and rejected claims by the victims. The government did not intercede on behalf of those affected. It took long and costly court battles to win partial victories. Without volunteer activists little progress would have been made. Is this the way it ought to be?

The victims of pollution in Japan were easily identified, the offenders less so. In the Reserve Mining Company episode, the reverse occurred. The taconite facility was an obvious polluter in some respects even though its more harmful emissions could not be detected with the naked eye and were not introduced as evidence until later in the trial. But there were no identifiable victims because the ill effects of asbestos take time to develop. And it would also be difficult to determine where around Lake Superior the contamination would be most serious in the long run. With so much guess work, was it right that an alarm was raised in the first place?

Our examples dealt with relatively local situations. Envisage, now, pollution which spreads widely. Acid rain in Scandinavia is ascribed to emissions from the Ruhr Valley in Germany. The buildup of carbon dioxide in the atmosphere and the resultant greenhouse effect, which could have adverse effects on the whole world, have their origins in the large amounts of fossil fuel used for energy by the industrial nations. Similarly, damage to the protective ozone layer of the earth's atmosphere through the release of freon is related to technological products used by the populations of those same nations. On the one hand one might expect the violation of borders to produce stronger countermeasures arising at the level of national govern- ments; on the other, the lack of focus on specific localities may diffuse the issues. As we remarked in Chap. 4, risks are not perceived as strongly when there is a lack of identification with the potential victims. Is it necessary, then, that we rely on environmental activists to refresh our awareness?

Any search for answers as to how the responsibility for environmental degradation should be shared will invariably lead us to further questions about how polluters and pollutants should be controlled. The first recourse recommended in all these matters is usually the judicial system. Let the courts decide who is guilty and how they are to be stopped! We have already described some of the shortcomings of this approach in Chap. 3. To reiterate: The courts move slowly, few individuals (whether plaintiff or defendant) can afford the process, and overreliance on the law promotes minimal compliance. In environmental cases the difficulties are compound- ed. A judge cannot be expected to be a specialist on health, safety, and the environment, nor can he or she usurp the powers of the legislature and prescribe control mechanisms. The most a judge can do is to guarantee a fair legal process.

Assuring a fair legal process is in itself a major undertaking. Technical information supportive of either the plaintiff or the defendant must be provided by expert testimony. Invariably both sides to a dispute will succeed in marshaling expert witnesses whose testimony will be contradic- tory. This should not surprise us because in spite of a professed adherence to rationality and commitment to truth, the engineer or scientist testifying will rely on a personal value system in selecting and presenting his or her information, mistrusting other interpretations. Earlier we had occasion to stress the importance of stating one's biases when serving as a legal witness or consultant. In an adversary hearing, admissions of bias and—perhaps more significantly—admissions of uncertainty can be exploited by the opposing side to such an extent that many specialists have become hesitant about giving evidence, to the detriment of all concerned. Rachel Carson, the embattled author of *Silent Spring,* described the consequences of this problem in a letter to a friend:

I'm convinced there is a psychological angle in all this that people, especially professional men, are uncomfortable about coming out against something, especially if they haven't absolute proof that "something" is wrong, but only a good suspicion. So they will go along with a program about which they privately have acute misgivings (quoted in Graham, 23).

Those experts who agree to testify open themselves to a barrage of hostile questions which can become personal in nature. Considerable embarrassment can occur if the truth is stretched or hidden. In the Reserve Mining Company trials some of the witnesses did not fare well. One expert who had served as a consultant to the plaintiffs later testified for the defendant, whom he could furnish with inside information. One expert claimed to have a Ph.D. degree from Purdue when in fact he did not. One consultant to Reserve was removed from a National Academy of Sciences committee studying the effects of asbestos in drinking water when it was disclosed he had agreed to provide Reserve with information on the confidential deliberations of the committee. There were conflicts of interest linking individuals and offices among the State of Minnesota, the University of Minnesota, Reserve, and the judges hearing the case. How believable can expert testimony be under such circumstances?

With so many uncertainties, one approach would be not to release a new product or process until it is shown to be risk-free. Industry and its engineers might claim that this is tantamount to declaring something to be guilty merely because innocence cannot be established. Is this not against our legal principles? Why does the U.S. Department of Agriculture declare poultry with 5 parts per million (ppm) or more of PCBs to be unfit for human consumption? The 5-ppm limit would be difficult to justify, just as a higher limit would be dangerous to defend, on strictly scientific grounds (Grundy, Weisbrod, and Epstein, 112). Where should the burden of proof lie? It is estimated that victims of Kanemi rice-oil disease had taken in at least 500 mg of PCBs. Translated into pounds of chicken contaminated with 5 ppm of PCBs, this would amount to 220 pounds of chicken or over 4 pounds of chicken consumed per week for a year. If this seems like a lot of chicken to eat, how much less would suffice to poison people with PCBs? In addressing the problem of where the burden of proof should lie, Garrett Hardin states:

In criminal law, as practiced in Britain and America, a man is "innocent until proven guilty. . . . " Scientists, however, see things otherwise. Science is an occupation in which most experiments fail. . . . Confronted with any new, untried, nostrum, a scientist, if called upon to place a bet, will bet that it won't work. Such is the conservative judgement (Hardin, 58–59).

Engineers have more confidence in their projects and will therefore chafe at such an interpretation. But they would have to be omniscient if they could foresee all the environmental effects of their work. And they would have to be omnipotent if they hoped to control those consequences without redesigns or retrofits. Accordingly, they must adopt the viewpoint of engineering as experimentation. The guilty-until-proven-innocent approach will then appear less unreasonable. Indeed it is not threatening since it applies to things and not to persons. If the expected benefit of a project or product is so great that prolonged testing seems an unreasonable cost, an earlier release could be authorized—provided that funds were available for subsequent corrective action, should it become necessary.

Internalizing Costs of Environmental Degradation

When we are told how efficient and cheap many of our products and processes are—from agriculture to the manufacture of plastics—the figures usually include only the direct costs of labor, raw materials, and use of facilities. If we are quoted a dollar figure, it is at best an approximation of the price. The true cost would have to include numerous indirect factors such as the effects of pollution, the depletion of energy and raw materials, and social costs. If these, or an approximation of them, were internalized— that is, added to the price—then those for whose benefit the environmental degradation had occurred could be charged directly for corrective actions. The problem with the "technofix" approach—using technology to repair the damages of technology—is not so much in physical realization as it is in financial burden. As taxpayers are beginning to revolt against higher levies, the method of having the user of a service or product pay for all its costs is gaining more favor. The engineer must join with the economist, the natural and physical scientists, the lawyer, and the politician in an effort to find acceptable mechanisms for pricing and releasing products so that the environment is protected through truly self-correcting procedures rather than adequate-appearing yet often circumventable laws. But again, as we did in our discussion on designing for safety in Chap. 4, we wish to point out here that good design practices may in themselves provide the answers for environmental protection without added real cost. For example, consider the case of a lathe which was designed to be vibration-free and manufactured to close tolerances. It not only met occupational safety and health standards with regard to noise, which its predecessor had not, but it also was more reliable, more efficient, and had a longer useful life, thus offsetting the additional costs in production (Melman, 1982, 176).

Technology Assessment

The Congress of the United States of America has an Office of Technology Assessment. It prepares studies on the social effects of technology in areas such as cashless trading (via bank card), nuclear war, or pollution. At the federal and state levels, many large projects must be examined in regard to their environmental impact before they are approved. The purpose of all this activity is praiseworthy. But how effective can it be?

Engineers, it is often said, tend to find the right answers to the wrong questions. The economist Robert Theobald made the following comment on education:

> The university is ideally designed to insure that you remain certain that you know the answers to questions that other people posed long ago. The problem today is that the questions we should be answering are not yet known. Unfortunately the process required for discovering the right questions is totally different from the process of discovering the right answers (quoted in Thrall and Starr, 17).

It should be quite apparent that it is not easy to know what questions to ask. And technology assessment and other forecasting methods suffer from this problem.

When scientists conduct experiments they endeavor to distill some key concepts out of their myriad observations. As shown in Fig. 7-1, a funnel can be used to portray this activity. At the narrow end of the funnel we have the current wisdom, the state of the art. Engineers make use of it to design and build their projects. These develop in many possible directions, as shown by the shape of the lower, inverted funnel. The difficult task of technology assessment and environmental impact analyses is to explore the

FIGURE 7-1
Distilling and applying knowledge.

Scientific inquiry

Distilled knowledge

Engineering applications

extent of this spread and to single out the more significant possible adverse effects.

The danger in any assessment of technology is that some serious risks can easily be overlooked while the studies and subsequent reports, properly authenticated by the aura of scientific methodology, assure the decision maker that nothing is amiss—or perhaps that perceived risks are more serious than they really are. We do not wish to belittle such efforts and we think that they are worthwhile, if only because of those questions and answers they raise which otherwise might not have surfaced. But there is a danger in believing that no further action is required once the reports have been approved and filed. Our contention remains that engineering must be understood as social experimentation and that the experiment continues, indeed enters a new phase, when the engineering project is implemented. Only by careful monitoring will it be possible to gather a more complete picture of the entangled web of effects encompassed in Fig. 7-1 within the inverted, lower funnel.

Study Questions

1 Describe how products or processes in an engineering field of interest to you can damage the environment, including human beings. Then examine pertinent laws designed to protect the environment at the local, state, and national levels and determine their effectiveness. Discuss alternatives to the regulatory approach.

2 Write an essay on the topic "Why Save Endangered Species?" You may wish to discuss some of the following points: Is it reasonable to put much effort into saving endangered species from extinction? What, if any, moral theories apply to plant and animal life and our relation thereto? Are we duty-bound to preserve all forms of natural life, or does such life enjoy an intrinsic right to exist? (Compare with Rescher, Essay 7; and with Regan.)

3 Are there any moral reasons why landowners may on occasion be constrained from doing to their properties what they please?

4 Phosphate detergents were alternately promoted and banned between 1947 and 1971. This caused great confusion in some communities (Lawless, 450). Critically examine the use of phosphate detergents as an experiment involving the environment.

5 The tragedy of an overused commons, first described by William Foster Lloyd in 1833 and retold by Garrett Hardin (Hardin, 254), paints the scenario of a village commons shared as a grazing ground by the villagers. One villager increases his herd by one animal but claims that the effect overall is as if the other villagers had each added only a fraction of an animal to their herds. Soon all the villagers behave this way and the pasture becomes overgrazed. Instead of maximizing their benefits, therefore, the villagers have lost all use of the commons. Draw on this paradigm in examining statements from, say, land developers and road builders that each of their projects will add only a minuscule amount of contamination to the environment.

6 Environmentalists are often accused of being elitists who wish to preserve the environment for their own enjoyment without regard for the needs of others. Granted that environmental controls may be necessary to preserve our habitat in the long run, do rich people (or nations) have the right to impose such controls now when they will harm poor people (or nations) more in the short run than they harm the rich?

7 Consider the following example of environmental side effects cited by Garrett Hardin:

> The Zambesi River . . . was dammed . . . to create the 1700-square-mile Lake Kariba. The effect *desired:* electricity. The "side-effects" *produced:* (1) destructive flooding of rich alluvial agricultural land above the dam; (2) uprooting of long-settled farmers from this land to be resettled on poorer hilly land that required farming practices with which they were not familiar; (3) impoverishment of these farmers . . . [and various other social disorders]; (6) creation of a new biotic zone along the lake shore that favored the multiplication of tsetse flies (Hardin, 68).

Similar problems have occurred when dams were built in the United States and when the Aswan Dam was erected on the Nile. One might ask if the original purpose may not itself begin to look like merely a side effect. If so, Hardin asks, can we *never* do anything? Describe under what conditions you think a dam such as the one on the Zambesi River should be built and operated. To whom is the engineer in charge of its construction ultimately responsible?

8 On January 15, 1919, a large wooden tank, 58 feet high and over 90 feet in diameter and holding over 2.5 million gallons of molasses, suddenly burst open. The immediate vicinity around this tank, which stood in the north end of Boston, was immediately covered by a stream of molasses which reached up to the second floor of many buildings. Numerous people were trapped by the sticky fluid and drawn down into it. What simple measure could have prevented the fluid from spreading and creating such an unusual "molasses environment?" Even today there are tanks holding vast quantities of water, oil, or liquefied natural gas where no adequate measures have been taken to protect the public from the dangers they present.

WEAPONS DEVELOPMENT

Much of the world's technological activity has a military focus. In terms of dollar (or equivalent currency) volume, direct or indirect involvement of engineers, and startling new developments alone it would deserve our serious attention. But the moral problems are of a magnitude which makes an inclusion of a discussion mandatory. The problems are not new to our times, but the consequences of nuclear war are. However, let us take a leap back into history first.

When Xerxes reached the Hellespont on his westward march, he had a bridge built across the waterway.

It is said that the bridge succumbed to a storm, the engineers were beheaded, the water was given a ceremonial flogging, and a second bridge was ordered. This time no chances were taken as 674 galleys in two rows were tied together with flax ropes weighing fifty pounds to the foot. The roadbed of planks, brush and earth withstood the crossing of 150,000 soldiers and several times that number of noncombatants (De Camp, 89).

There are several reasons for an engineer to do his or her best on a military job. High among them are patriotism and prudential interest. The latter can be occasioned by threats from an easily displeased ruler or by the lure of commercial success. Xerxes' bridge builders certainly had compelling reasons to build a good bridge. With their experience they might even have benefited from staying on at the site, offering to build bridges for other armies who came along. (Indeed, the next that came along was Xerxes' army on retreat.)

The Weapons Seesaw

The trade in arms and military know-how has an honored tradition. Today military expenditures throughout the world total about $500 billion dollars annually. Of this amount, one-quarter is earmarked for purchases of weapons and related equipment, 17 percent of which are traded internationally.

Among the world's most successful arms merchants and manufacturers was the family Krupp. Krupp was an expert at exploiting the defensive/offensive weapons seesaw. This is how it worked, according to William Manchester in his book *The Arms of Krupp:*

> Having perfected his nickel steel armor, Fritz advertised it in every chancellery. Armies and navies invested in it. Then he unveiled chrome steel shells that would pierce the nickel steel. Armies and navies invested again. Next—this was at the Chicago fair, and was enough in itself to justify the pavilion—he appeared with a high-carbon armor plate that would resist the new shells. Orders poured in. But just when every general and admiral thought he had equipped his forces with invincible shields Fritz popped up again. Good news for the valiant advocates of attack: it turned out that the improved plate could be pierced by "capped shot" with explosive noses, which cost like the devil. The governments of the world dug deep into their exchequers, and they went right on digging. Altogether thirty of them had been caught in the lash and counterlash. Fritz showed the figures to the emperor, who chuckled. He would have gagged if he had known the truth: he himself was being trapped in a variation of the seesaw (Manchester, 248).

In the Far East the firm of Krupp (Germany) was joined by Vickers (Great Britain) and Schneider (France) in supplying arms to the Chinese, the Japanese, and the Russians. Sir William White, chief designer of the British Admiralty and later director of warship construction at Armstrong, Whitworth and Co., was said not to be unwilling

> to play the part of the *honnete coutier* by pointing out the growth of the Japanese navy to his Chinese clients, or of the Chinese to their indomitable rivals. . . . By such means he was able to increase the profits of the great company which employed him, and to extend what is, perhaps, the most important of our national industries, and to kindle in the hearts of two Asiatic peoples the flames of an enlightened and sacred patriotism (Seldes, 36).

Everything worked out quite well for the arms merchants when after both the Japanese and Russian fleets had been outfitted, the latter was badly battered during the Russo-Japanese War in 1905, requiring new orders to rehabilitate it.

While Krupp men were busy with their steel, Krupp wives did good deeds among the workers' families. This was good stuff for satire, superbly used in George Bernard Shaw's portrayal of the Krupps in *Major Barbara*. Berta Krupp once expressed horror at the fact that the Essen Works would manufacture "bomb cannons." She is said to have been assured that few would dare get into the way of such a cannon—thus it would promote peace. By World War I, heavy ordnance took the form of 420-millimeter howitzers lofting 1-yard-long shells weighing 1 ton each over 9 miles, and of 210-millimeter cannons which fired 264-pound shells at Paris from 76 miles away. Both types of guns were occasionally dubbed Dicke (Big) Berta, some say in honor of Berta Krupp, others say in memory of the monk Berthold Schwarz, the reinventor of gun powder in the western world.

Big Berta was superseded in World War II by Big Dora, two stories high with a barrel 90 feet long and a bore of 840 millimeters. This monster could fire 8-ton shells up to 30 miles. But by then the big guns were no longer as useful. Mobile delivery of destructive force by tanks and aircraft had been perfected, to be followed by rockets and nuclear weapons.

This change in technology was accompanied by a change in strategy. Where armies once faced each other in battle, bombs could now be delivered far beyond the front lines on the enemy's population centers. The scene of battle was thus moved, and the change is reflected in the work of Erich Maria Remarque, whose *All Quiet on the Western Front* dealt with the sufferings of the soldier in the front lines of World War I, but whose World War II novel *A Time to Live, a Time to Die* centered on the homefront and dealt with the effects of war on the civilian population.

In the early months of World War II the chiefs of staff of the warring countries would still agonize over the question of whether or not to bomb targets in civilian population centers at night. Toward the end of the war, night raids had become common practice and civilians themselves had become the targets. The agonizing decisions were now up to them: Which child or children do you leave behind if the asphalt roads are so soft from the heat that you cannot possibly escape the flames if you carry all?

To us the atom bombs dropped on Hiroshima and Nagasaki are horrible not only because of the many deaths they caused (single air raids during World War II had killed larger numbers of people), nor because of the ghastly medical consequences for survivors, but mostly because they ushered in the age of rapid, irretrievable delivery of destructive power in immense concentrations. To deliver the equivalent explosive effect of the Hiroshima bomb would have required 20,000 tons of explosives (TNT). Loaded on railroad hopper cars this would have filled 267 cars making a train 2 miles long, just for one bomb. It would have taken 740 of today's B-52 bombers to carry this load in conventional explosives.

Twenty years ago the Soviet Union exploded hydrogen bombs in the 50- and 60-megaton range for test purposes. Each of those bombs had two to three thousand times the destructive power of the Hiroshima bomb, and it would take a train 4000 to 6000 miles long to carry the TNT required to produce an explosion equivalent to that produced by any one of them. You would have to wait at a railroad crossing for close to 100 hours to let such a train pass. Or if carried by bombers, you would see the sky darkened with one and a half million planes. The total megatonnage of nuclear weapons available to the two major superpowers for instant departure toward targets stood at about 7700 in 1980, with 4500 megatons available to the USSR and 3200 megatons to the United States (Stockholm International Peace Res. Inst., pp. XXVII–XXVIII).

Today's bombs are smaller (though often clustered), mostly carried by missiles, and targetable with great exactness. Countermeasures are being developed and the offensive/defensive seesaw is exercised in the same manner as a century ago. The Office of Technology Assessment of the United States Government estimates that a large-scale nuclear exchange between the United States and the USSR could kill more than 250 million people outright in those two countries alone, let alone those who would suffer lingering illnesses and later deaths from starvation, exposure, or disease.

Those consequences are not disputed. What occupies the best minds are questions regarding national policy: Is the threat of mutual annihilation an effective deterrent for the future? Would unilateral disarmament, or disarmament without adequate safeguards, be an effective indicator of trust which could bring about a deescalation of the arms race and the potential for nuclear war? In the meantime the nuclear arsenal becomes bigger and more technically perfect.

The Engineer's Involvement in Weapons Work

Instruments of torture leave little to the imagination. Descriptions of the use of the rack, the thumbscrew, and the electric prod convey instant sensations of pain and suffering. Our modern weapons of war, however, at first seem more remote. Unless one has experienced or studied their consequences, they appear quite guiltless in themselves. Even those who deliver their deathly charges against the enemy mostly do so at a distance, making their use seem more acceptable—or at least less unsettling.

How do the men and women who design weapons, manufacture them, them, and use them feel about their work? Most have reservations about it, but those who stay in it also have reasons to support their continued involvement. The following cases involve real weapons. The people are composites who represent the various positions taken by typical citizens who are also engineers.

1 Bob's employer manufactures antipersonnel bombs. By clustering 665 guava-sized bomblets and letting them explode above ground, an area covering the equivalent of ten football fields is subjected to a shower of sharp fragments. Alternatively the bombs can be timed to explode hours apart after delivery. Originally the fragments were made of steel; now less easily detected plastics are sometimes used, making the treatment of wounds, including the location and removal of the fragments, more time-consuming for the surgeon. Recently another innovation was introduced: By coating the bomblets with phosphorus the fragments could inflict internal burns as well. Thus the antipersonnel bomb does its job quite well without necessarily killing in that it ties up much of the enemy's resources in treating the wounded who survive its explosion. Bob himself does not handle the bombs in any way, but as an industrial engineer he enables the factory to run efficiently. He does not like to be involved in making weapons, but then he tells himself that someone has to produce them. If he does not do his job, someone else will, so nothing would change. Furthermore, with the cost of living being what it is, he owes his family a steady income.

2 Mary is a chemical engineer. A promotion has gotten her into napalm manufacturing. She knows it is nasty stuff. She remembers Professor Wald, a Nobel laureate in biology from Harvard, berating the chemical industry for producing this "most brutal and destructive weapon that has ever been created". But this was when she was in college, during the Vietnam war. Civilians were forever not leaving the fighting zone and then there were complaints about them being hurt or killed. She abhors war like most human beings, but she feels that the government knows more than she does about international dangers and that it is better to fight a war abroad than on our shores. If everyone were to decide on her or his own what to do and what not to do, then there would be utter chaos. Perhaps society can tolerate a few oddballs with their own ideas, but companies certainly should be prepared to manufacture the weapons our armed forces need. Incidentally, if Mary continues to perform well on her job she will be promoted out of her present position into working on a commercial product with much growth potential.

3 Ron is a specialist in missile control and guidance. He is proud to be able to help his country through his efforts in the defense industry. The missiles he works on will carry single or multiple warheads with the kind of dreadful fire power which, in his estimation, has kept any potential enemy in check since 1945. At least there has not been another world war—the result of mutual deterrence, according to his point of view.

4 Marco's foremost love is physical electronics. He works in one of the finest laser laboratories. Some of his colleagues do exciting research in particle beams. That the laboratory is interested in developing something akin to the "death ray" described by science fiction writers of his youth is of secondary importance. More bothersome is the secrecy which prevents him from freely exchanging ideas with experts across the world. But why change jobs if he will never find facilities like those he has now?

5 Joanne is an electronics engineer whose work assignment includes avionics for fighter planes which are mostly sold abroad. She has no qualms about such planes going to what she considers friendly countries, but she draws the line at their sale to potentially hostile nations. Joanne realizes that she has no leverage within the company, so she occasionally alerts journalist friends with news she feels all citizens should have. "Let the voters direct the country at election time"—that is her motto.

6 Ted's background and advanced degrees in engineering physics gave him a ready entry into nuclear bomb development. As a well-informed citizen he is seriously concerned with the dangers of the ever-growing nuclear arsenal. He is also aware of the possibilities of an accidental nuclear exchange. In the meantime he is working hard to reduce the risk of accidents such as the thirty-two "broken arrows" (near catastrophic accidents) reported by the Pentagon—or the many others that he knows have occurred worldwide. Ted continues in his work because he believes that only specialists, with firsthand experience of what modern weapons can do, can eventually turn around the suicidal trend represented by their development. Who else can engage in meaningful arms-control negotiations?

Our names are fictitious, as are the specific jobs described. But all over the world talented people engage in weapons work. Surely everyone who accepts a job in a war-related industry should also seriously consider his or her motives in doing so. Prudential self-interest is not sufficient to guarantee responsible participation in what must be regarded as mankind's most crucial engineering experiment. Only those who have arrived at morally autonomous, well-reasoned positions for either engaging in or abstaining from weapons work can be counted on to carefully monitor the experiment and try not to let it run a wild course.

Defense Industry Problems

Compared to the major problem of war and peace, the moral issues which arise in connection with defense contracts must appear trifling—as indeed might all of the issues presented in the earlier chapters. But we must concern ourselves with the microscopic as well as the macroscopic context because the two are linked. It is one world we inhabit, and we carry our attitudes and convictions from one part of our lives to another. Accordingly what goes on at the work place and what we learn from the daily local news color our views about larger issues and eventually influence how we relate to them.

Defense-related work has always been a fertile ground for unethical behavior by unscrupulous contractors. As is the case in many public-works projects, there is first of all the matter of size. Large contracts involve large sums of money, they are difficult to oversee, and bureaucracy hinders even the most conscientious overseer. This is compounded in military projects by secrecy imposed for national security purposes. Finally there is the problem of urgency: readying a weapons system before it is obsolete. On a different level, there is the problem of "technology creep"—the development of new weapons which alters diplomatic arrangements, even while they are being negotiated.

The following cases illustrate the various problems just enumerated and should give pause for ethical reflection by both critics and ardent supporters of weapons manufacturing and development.

1 The problem of waste and cost overruns is a continuing one in the defense industry (E. Fitzgerald; Gansler; Melman). In Chap. 6 we mentioned one example—the $2 billion cost overruns in developing the C5-A transport plane, overruns reported to the public by Ernest Fitzgerald. Fitzgerald, who is a deputy for management systems in the Pentagon, has been a significant critic of how the defense industry has operated at efficiencies far below commercial standards. He has described how contractors' work forces were swelled with underutilized engineers and high-salary sales personnel, resulting in lavish overhead fees. Or how small contractors were willing to comply with cost-cutting plans, but large suppliers felt secure in not complying. (At present, twenty-five firms hold 50 percent of all defense contracts in the United States, and eight firms conduct 45 percent of all defense research.)

High cost and poor quality were encouraged in various ways: Planned funding levels were leaked to prospective contractors. Cost estimates were based on historical data, thus incorporating past inefficiencies. Costs were cut when necessary by lowering quality, especially when component specifications were not finalized until the contract was completed. Sole-supplier policies gave a contractor the incentive to "buy in" with an artificially low bid, only to have the contractor plead for additional funds later on. And those funds were usually forthcoming, since the Department of Defense has historically accepted what it knows to be optimistically low development-cost estimates because they stand a better chance of being approved by Congress (Gansler, 296).

2 In Goethe's poem *Der Zauberlehrling*, the sorcerer's apprentice employs his master's magic incantation to make the broom fetch water. When he cannot remember the proper command to stop the helpful broom, however, he comes near to drowning before the master returns. Technology resembles the broom as the slave in service of the arms race. Not only has the world's arsenal grown inordinately expensive (even without graft), and not only does it contribute to a steadily worsening inflation, it has also gained a momentum all its own. George Kennan, a former ambassador to the Soviet Union, has stated:

> I see this competitive buildup of armaments conceived initially as a means to an end but soon becoming the end itself. I see it taking possession of men's imagination and behavior, becoming a force in its own right detaching itself from the political differences that initially inspired it, and then leading both parties, invariably and inexorably, to the war they no longer know how to avoid (Kennan).

The arsenal is not only growing in size, it is also getting "better." Diplomats may be striving to avert a major conflict, but all the while an exuberance for new developments creates a technology creep which can at times postpone or even upset all negotiations. Nations are suddenly seen to shift to new positions as a new device to more accurately target missiles, or perhaps an entirely new weapon, are reported to be in the offing. This can destabilize (and occasionally stabilize) the political process. Meanwhile, as Jerome Wiesner and Herbert York have written:

> . . . both sides in the arms race are . . . confronted by the dilemma of steadily decreasing national security. *It is our considered professional judgment that this dilemma has no technical solution.* If the great powers continue to look for solutions in the area of science and technology only, the result will be to worsen the situation (Wiesner and York, 27).

The technological imperative that innovations must be implemented should give advocates of conventional, limited war preparedness some cause for concern as well. Giving in to the excitement of equipping and trying out weapons employing the latest in technology may provide added capability to sophisticated, fully automatic systems such as intercontinental ballistic missiles. But if tactical, humanly operated weapons fall prey to the gadget craze, a less than optimal system may result. The F-15 fighter illustrates this problem of preoccupation with prestige-boosting modernism. The plane is the fastest and most maneuverable of its kind, yet 40 percent of the F-15s are not available for service at any one time because of defects, difficulty of repair, and lack of spare parts. We observe that faddishness is not restricted to the world of fashion—it occurs in technology as well. The engineer must constantly be on guard not to fall prey to it.

3 The final example concerns peacetime secrecy in work of military import. Norbert Wiener, the founder of cybernetics, was moved to write in answer to a request for information on some of his wartime work in missile control, "that to provide information is not necessarily an innocent act." He refused to give the information because too often the scientist places "unlimited powers in the hands of people he is least inclined to trust with their use" (Wiener, 1947, 46).

Secrecy poses problems for engineers in various ways. Should discoveries of military significance always be made available to the government? Can they be shared with other researchers, with other countries? Or should they be withheld from the larger scientific and public community altogether? If governmental secrecy in weapons development is allowed to become all-pervasive, on the other hand, will it also serve to mask corruption or embarrassing mistakes within the defense establishment? There are no easy answers to these questions, but it should be no surprise that universities, as institutions committed to the free exchange of ideas, generally shun secret research projects.

Study Questions

1 Is it right to ask whether a weapons system is cost-effective? What does it matter how expensive it is when the nation's security is at stake and the weapons work as intended?

2 In a farewell address to the nation in 1961, outgoing President Dwight Eisenhower sounded a warning of the emerging military-industrial complex. He told his listeners that we cannot immediately tool up for massive weapons production should another major war occur, that our arms must thus be ready and mighty, and that therefore we must have a large peacetime defense establishment. But he went on to say:

This conjunction of an immense military establishment and a large arms industry is new in the American experience. The total influence, political, even spiritual, is felt in every city, every State house, every office of the Federal Government. We recognize the imperative need for this development. Yet we must not fail to comprehend its grave implications. Our toil, resources and livelihood are all involved; so is the very structure of our society.

What implications did Eisenhower have in mind? Are there ethical implications as well? Of what kind?

3 Earl Louis Mountbatten spoke the following words shortly before his assassination:

Next month I enter my eightieth year. I am one of the few survivors of the First World War who rose to high command in the second and I know how impossible it is to pursue military operations in accordance with fixed plans and agreements. In warfare the unexpected is the rule and no one can anticipate what an opponent's reaction will be to the unexpected. . . . There are powerful voices around the world who still give credence to the old Roman precept—if you desire peace, prepare for war. This is absolute nuclear nonsense and I repeat—it is a disastrous misconception to believe that by increasing the total uncertainty one increases one's own certainty (address on the occasion of being awarded the Louise Weiss Foundation prize; delivered at Strasbourg in 1979 to the Stockholm International Peace Research Institute).

Interpret Mountbatten's words from the viewpoint which sees modern war as a technological experiment.

4 Consider the following blunt pronouncements by statesmen and generals (Schwartz, 203). Is there any message in them for engineers?
  a Douglas MacArthur (commencement speech at Michigan State University, 1961): "Global war has become a Frankenstein to destroy both sides. If you lose you are annihilated. If you win, you stand only to lose. No longer does it possess even the chance of the winner of a duel. It contains only the germs of double suicide."
  b Dwight D. Eisenhower, after the collapse of the 1960 summit meeting: "All of us know that, whether started deliberately or accidentally, global war would leave civilization in a shambles. In a nuclear war there can be no victors—only losers."
  c Nikita Khrushchev, when Premier of the USSR: "Only madmen and lunatics can now call for another world war. As for the men of sound mind—and they account for the majority even among the most deadly enemies of communism —they cannot but be aware of the fatal consequences of another war."
  d John F. Kennedy, addressing the United Nations: "The weapons of war must be abolished before they abolish us."
5 The Just-War theory considers a war to be acceptable when it satisfies several stringent criteria: The war must be fought for a just *cause,* the *motives* must be good, it must follow a call from higher *authority* to legitimize it, and the use of *force* must be based on necessity (Hehir, 368–369). Central to notions of a just war are the principles of noncombatant immunity and proportionality. Noncombatants are those who will not be actively participating in combat and therefore do not need to be killed or restrained. Proportionality addresses the extent of damage or consequences allowable in terms of need and cost. Describe a scenario for the

conduct of a just war and describe the kinds of weapons engineers might have to develop to wage one.

6 The following problem is taken from an article by Tekla Perry in the *IEEE Spectrum* [although it involves the National Aeronautics and Space Administration rather than the Defense Department, many of the actors (companies and government) involved in space research are also involved in weapons development]:

Arthur is chief engineer in a components house. As such, he sits in meetings concerning bidding on contracts. At one such meeting between top company executives and the National Aeronautics and Space Administration, which is interested in letting a major contract, NASA presents specifications for components that are to be several orders of magnitude more reliable than the current state of the art. The components are not part of a life-support system, yet are critical for the success of several planned experiments. Arthur does not believe such reliability can be achieved by his company or any other, and he knows the executives feel the same. Nevertheless, the executives indicate an interest to bid on the contract without questioning the specifications. Arthur discusses the matter privately with the executives and recommends that they review the seemingly technical impossibility with NASA and try to amend the contract. The executives say that they intend, if they win the contract, to argue mid-stream for a change. They remind Arthur that if they don't win the contract, several engineers in Arthur's division will have to be laid off. Arthur is well-liked by his employees and fears the lay-offs would affect some close friendships. What should Arthur do? (Perry, 58).

After you have prepared your answer, you might wish to consult the source of this case for various other reactions to the dilemma it presents.

CONSULTING ENGINEERING

If we have succeeded in making the preceding section on military-related work a soul-searching exercise regarding our widest perceptions of humankind and its place in the world, then the present section on consulting engineering should serve to focus our attention again on the inescapable, day-to-day problems of the independent engineer. Both exercises are important because we live in a world where both the macroscopic and the microscopic aspects of life take their rightful place.

Consulting engineers operate in private practice. They are paid by fees for the services they render, and not by salaries received from employers. Because of this, they tend to have greater freedom to make decisions about the projects they undertake. Yet their freedom is not absolute: They share with salaried engineers the need to earn a living.

Here we will discuss four topics—advertising, competitive bidding, contingency fees, and provisions for resolution of disputes—which illustrate some of the areas in which consulting engineers have special responsibilities. We will also note how in safety matters consulting engineers may have greater responsibility than salaried engineers, corresponding to their greater freedom.

Advertising

Some corporate engineers are involved in advertising because they work in product sales divisions. But within corporations, advertising of services, job openings, and the corporate image are left primarily to advertising executives and the personnel department. By contrast, consulting engineers are directly responsible for advertising their services, even when they hire consultants to help them.

Prior to a 1976 Supreme Court decision, competitive advertising in engineering was considered a moral issue and was banned by professional codes of ethics. As in law and medicine, anything beyond a tasteful notification of the availability of one's services was thought to be "unprofessional." It was deemed unfair to colleagues to win work through one's skill as an advertiser rather than through one's earned reputation as an engineer. It was also felt that competitive advertising caused friction among those in the field and lessened their mutual respect and that vigorous advertising damaged the profession's public image by placing engineering on a par with purely money-centered businesses.

However, the Supreme Court disagreed with that view. According to its ruling, as well as other rulings by the Federal Trade Commission, general bans on professional advertising are improper restraints of competition. They serve to keep prices for services higher than they might otherwise be, and they also reduce public awareness of the range of professional services available.

These rulings have shifted attention away from whether professional advertising is moral or not toward whether it is honest or not. Deceptive advertising normally occurs when products or services are made to look better than they actually are. This can be done in many ways: (1) by outright lies, (2) by half-truths, (3) through exaggeration, (4) by making false innuendos, suggestions, or implications, (5) through obfuscation created by ambiguity, vagueness, or incoherence, (6) through subliminal manipulation of the unconscious, etc. (Leiser).

There are notorious difficulties in determining whether specific ads are deceptive or not. Clearly it is deceptive for a consulting firm to claim in a brochure that it played a major role in a well-known project when it actually played a very minor role. But suppose the firm makes no such claim and merely shows a picture of a major construction project in which it played only a minor role? Or, more interestingly, suppose it shows the picture along with a footnote which states in fine print the true details about its minor role in the project? What if the statement is printed in larger type and not buried in a footnote?

As another example, think of a photograph of an engineering product (say, an electronics component) used in an ad to convey the impression that the item is routinely manufactured and available for purchase, perhaps even "off the shelf," when in actuality the picture shows only a preliminary prototype or mockup and the item is just being developed. To what extent, then, should the buyer—as the subject or participant of an "experiment" conducted by the manufacturer—be protected from misleading information about a product?

Advertisers of consumer products are generally allowed to suppress negative aspects of the items they are promoting and even to engage in some degree of exaggeration or "puffery" of the positive aspects. Notable exceptions are ads for cigarettes and saccharin products, which by law must carry health warnings. By contrast, norms concerning the advertising of professional services are much stricter. For example, the code of the National Society of Professional Engineers (NSPE) forbids all of the following:

> . . . the use of statements containing a material misrepresentation of fact or omitting a material fact necessary to keep the statement from being misleading; statements intended or likely to create an unjustified expectation; statements containing prediction of future success; statements containing an opinion as to the quality of the Engineer's services; or statements intended or likely to attract clients by the use of showmanship, puffery, or self-laudation, including the use of slogans, jingles, or sensational language format (*NSPE Code of Ethics*, Section 3b).

Is there sufficient warrant for these tougher restrictions? Do they perhaps suppress vigorous competitive advertising which could be beneficial to the public? How are reasonable restrictions on the manner of advertising to be justified in terms of ethical theories? These are questions which deserve to be addressed within engineering ethics.

Active solicitation of clients through advertising or personal contacts has been considered especially unprofessional when it takes work away from other engineers. Yet here again one may ask whether some degree of solicitation serves the public interest by encouraging healthy competition. The dangers of allowing it, of course, must also be weighed carefully. Does it open the door to those who are dishonest and who might in very subtle ways unfairly criticize the work of other engineers whom they seek to supplant, or to those who might exaggerate the merits of their work? Certainly strong restrictions on misleading advertising in this area are especially important.

Competitive Price-Bidding

For many years codes prohibited consulting engineers from engaging in competitive bidding, that is, from competing for jobs on the basis of submitting proposed fees. The following statement, for example, formerly appeared in the code of the American Society of Civil Engineers:

> It shall be considered unprofessional and inconsistent with honorable and dignified conduct and contrary to the public interest for any member of the American Society of Civil Engineers to invite or submit priced proposals under conditions that constitute price competition for professional services (quoted in Alger, Christensen, and Olmsted, 35).

It was considered permissible for industrial and construction firms to use competitive bidding because they could formulate cost estimates with some accuracy based on fixed design specifications. By contrast, the job of the consulting engineer is generally to develop creative designs for solving novel problems. Often there is no way to make precise bids. Allowing competitive bidding in such cases, it was felt, would open the door to irresponsible engineering in that inaccurate bids would encourage either cutting safety and quality (in the case of low bids) or padding and overdesigning (in the case of high bids).

However, in 1978 the Supreme Court ruled that professional societies were unfairly restraining free trade by banning competitive bidding. The ruling still left several loopholes, though. In particular, it allowed state registration boards to retain their bans on competitive bidding by registered engineers. It also allowed individual consulting firms to refuse to engage in competitive bidding. Thus fee competition where creative design is involved has remained a lively ethical issue. Is it in the best interests of clients and the public to encourage the practice?

If the use of competitive bidding is widely rejected by engineering firms, clients will have to rely almost exclusively on reputation and proven qualifications in choosing among them. This raises the problem of how the qualifications are to be determined in an equitable way. Is the younger, but still competent, consulting engineer placed at an unfair disadvantage? Or is it reasonable to view this disadvantage as justifiable, given the general importance of experience in consulting work?

Contingency Fees

Consulting engineers play the primary role in making arrangements about payment for their work. Naturally this calls for exercising a sense of honesty and fairness. But what is involved specifically?

As one illustration of the kinds of problems which may arise, consider the following entry in the code of the National Society of Professional Engineers:

> An Engineer shall not request, propose, or accept a professional commission on a contingent basis under circumstances in which his professional judgment may be compromised, or when a contingency provision is used as a device for promoting or securing a professional commission (NSPE Code of Ethics, Section 11b).

A "contingency" fee or commission is one which is dependent upon some special condition beyond the normal performance of satisfactory work. Typically a contingency-fee arrangement requires that the consultant is to be paid only if she or he succeeds in saving the client money. Thus a client may hire a consultant to uncover cost-saving methods which will save 10 percent on an already contracted project. If the consultant does not succeed in doing so, no fee is paid. The fee may be either an agreed-upon amount or a fixed percentage of the savings to be realized.

In many contingency-fee situations it is easy for the consultant's judgment to become biased. For example, the prospects of winning the fee may tempt the consultant to specify inferior materials or design concepts in order to cut construction costs. Hence the point of the NSPE code entry. But even allowing for this problem, is the thoroughgoing ban on such fees in the NSPE code warranted? There is, after all, a point to their use. They are intended to help stimulate imaginative and hopefully responsible ways of saving costs to clients or the public, and presumably this consideration deserves some weight.

Resolving the issue calls for balancing out the potential gains against the potential losses of allowing or banning the practice. In this respect it is like many other issues in engineering ethics which call for reasonable judgments based on both past experience and foresight. And philosophical ethical theories can be useful in making those judgments by providing a general framework for assessing the morally relevant features of the problems under consideration.

Safety and Client Needs

The greater amount of job freedom enjoyed by consulting engineers as opposed to salaried engineers leads to wider areas of responsible decision making concerning safety. It also generates special difficulties.

Very often, for example, consulting engineers have the option of accepting or not accepting "design-only projects." A design-only project is one where the consultant contracts only to design something, but not to have any role—even a supervisory one—in its construction. Design-only projects are sometimes problematic because of difficulties encountered in implementing the designing engineer's specifications and because that engineer is often the only individual really well qualified to identify the areas of difficulty (Alderman and Schultz, 8–12). For example, clients or contractors may lack adequately trained inspectors of their own. In fact, when novel projects are being undertaken, clients may not even know that their own inspectors are unsatisfactory. Again, a contractor may be unable or unwilling to spot areas where the original design needs to be modified so as to best serve the client. The designer is often the person best able to ensure that the client's needs are met, as well as the safety needs of the project, yet he or she may not be around to do so.

The importance of having the designer involved in on-site inspection is illustrated by the following example:

> An engineering firm designed a flood control project for the temporary retention of storm water in a nearby city. Included in the project were some high reinforced concrete retaining walls to support the earth at the sides of the retention basin. Although the consulting engineer had no responsibility for site visits or inspection, one of the designers decided to visit the site, using part of his lunch hour to see how it was progressing. He found that the retaining wall footings had been poured and the wall forms were placed. He was shocked, however, to find that the reinforcing steel extended from the footings into the walls only a small fraction of the specified distance. He immediately returned to his office and the client was notified of the situation. The inspectors responsible were disciplined and corrective measures taken with respect to the steel reinforcement. There is no question that in the first heavy rain the walls would have collapsed had the designer not discovered this fault. The result would have been heavy property loss, waste of resources, environmental damage and possible injury, even loss of life (Alderman and Schultz, 11).

It is thus a significant area of inquiry to determine when consulting engineers should or should not accept design-only projects. And when they do accept them, are they not obligated to make at least occasional on-site inspections later in order "to monitor the experiment" they have set in motion? That is, are there at least some minimal moral responsibilities in this context which reach beyond the legal responsibilities specified in the contract?

In the course of making on-site inspections consulting engineers may notice unsafe practices which endanger workers. For example, they may notice the absence of a sufficient number of secondary support struts for a building or bridge. They may know from their past experience that this could cause a partial collapse of the structure while construction workers are on the job. Of course job safety is the primary responsibility of the contractor who has direct control over the construction. Yet for the consultant to do nothing would be negligent, if not calloused. But how far do the consultant's responsibilities extend? Is a letter to the construction supervisor sufficient? Or is the consultant morally required to follow through by checking to see that the problem is corrected? It should be noted that an engineer who does point out construction deficiencies on one occasion— even if not contractually required to—but refrains from doing so on other occasions can be held liable for complicity in any damages resulting from unreported deficiencies. (Review study question 2 at the end of Chap. 4.)

Provision for Resolution of Disputes

Large and complex engineering projects involve many participants at different levels of responsibility within the organizations representing the owner, the consulting engineer, and the construction firm. Overlapping responsibilities, fragmented control, indecision, delays, and an inability to resolve disputes quickly and amicably characterize many projects. To forestall potential liabilities in such situations, the various parties involved usually devote much time to protecting themselves when it could more profitably be used to improve the quality of the project. Resolution of disagreements is made more difficult when construction lasts several years and personnel changes occur during that period, because mutual trust and understanding are not easily nurtured under such conditions.

It has been observed by engineers engaged in construction projects that

> . . . owners often initiate polarization from the outset by placing liability on engineers, vendors, and constructors without categorization or regard to which, or under what conditions faults may arise. . . . The engineer extends the chain by preparing tighter specifications, employing every exculpatory phrase at his command, and inviting the constructor to nominate his best "sea-lawyer" as project manager. All this leads to the compounding of heavy contingency factors—where possible—or further assumption of risk by the implementing entities (P. Smith, 33).

The position of owners is not difficult to understand, as they have the most to lose. Thus the tendency to shift risks onto others. Engineers are tied to the contract provisions. They will be reluctant to innovate, preferring instead to stick to the tried-and-true ways of doing things. Contractors are used to taking large risks, but they are quick to recognize conditions "beyond their control."

Litigation has increased considerably in recent decades, and the character of litigation has changed as well. The construction industry is no exception, and its experience, as described by a panel of experts, serves as a good illustration of the kinds of legal problems consulting engineers now sometimes face:

> Traditionally law suits were fairly clear cut and involved matters directly related to the construction process; suits by owners were relatively uncommon; the design professional had to contend with virtually no litigation; and disputes were almost entirely confined to participants in the construction process.

Today not only has the number of lawsuits dramatically increased but the nature of the lawsuits and the participants also have changed. Third forces, historically external to the process, today are the motivating factors behind a great many suits (Buehler, 94).

The third forces mentioned above typically are citizen intervenors or regulatory commissions.

Since litigation is time-consuming and costly, the consulting engineer should arrange contractually for methods of resolving conflicts. Quite apart from defining how risks are to be apportioned and payment of fees to be made, there should be contractual provisions for dispute-solving vehicles (designed to avoid costly court battles) such as mediation-arbitration in which a mediator attempts to resolve a dispute first, and if that is not fruitful, to act as the final, binding arbitrator. It should also be specified contractually that the National Joint Board for Settlement of Jurisdictional Disputes will be called upon to provide a hearing board and appeals board.

Engineering practice today does not regularly provide for clear-cut arbitration or conflict resolution. We believe that the consulting engineer is the proper party to assume the obligation of assuring that such clauses are included in contracts and are adhered to by everyone involved. This responsibility arises from the engineer's close contacts with both the owner and the constructor of a project and from the "social experimentation" nature of engineering. While contractual arrangements cannot solve problems caused by third-party intervenors, they can forestall a great many conflicts the origins of which can easily be foreseen and provided for.

Study Questions

1 Locate three advertisements: one for a technological product; one for the services of a consulting engineer; and one for positions in an engineering firm. Critique each ad in terms of whether the information or pictures included are misleading or deceptive in any way (be specific).
2 State why you agree, or why you disagree, with the following positions regarding advertising in local telephone directories. The first three examples comprise Case No. 72–1 of the *NSPE: Opinions of the Board of Ethical Review* (Vol. IV, 1976). The fourth is our adaptation of a case involving a dentist and the dentists' association in San Francisco.
  a "Are bold face listings in the classified section of local telephone directories consistent with the Code of Ethics?" The NSPE says "no."
  b "Are bold face listings in the regular section of local telephone directories consistent with the Code of Ethics?" The NSPE says "yes."
  c "Are professional card-type listings (set off by lines or blank space) in the classified section of local telephone directories consistent with the Code of Ethics?" The NSPE says "no."
  d Mr. Zebra is a consulting engineer whose firm, Zebra Associates, appears last in the telephone directory's classified listing of engineers. In order to gain a more

advantageous position in the yellow pages and in other directories, he changes the name of his firm to Antelope and Zebra. Antelope is a purely fictitious partner. Is this ethical? In a similar case a dentists' board of review said "no."

In considering the first three cases you may wish to consult the reasoning of the NSPE Board of Ethical Review. Also, examine some of the other numerous rulings by that board.

3 Is there anything unethical about the conduct of the engineer employee described in the following?

A firm in private practice handles many small projects for an industrial client, averaging 20 to 30 projects a year. The firm has a signed agreement with the industrial client which does not obligate the client to give the firm any work, but does establish the respective responsibilities, terms of payment and other contractual details when the client does use the firm's services. The actual assignments are made by means of purchase orders referring to the agreement. An engineer employee of the firm resigns his employment and establishes his own firm and then actively solicits the industrial client of his former employer without any prior indication of interest by the client (*NSPE: Opinions of the Board of Ethical Review,* Case No. 73–7, Vol. IV, 1976).

4 Is the decision of the consulting firm in the following example the morally obligatory one?

A large civil engineering consulting firm completed a comprehensive arterial highway plan for a large, midwestern metropolitan area. The area appropriated funds for the next phase of the program: the preparation of a design report covering the recommended highways. The officials concerned, not knowing what a reasonable fee would be for the design report work, felt that it was necessary to invite proposals from various consultants. The consulting firm explained that it could not participate if the selection were to be made on the basis of engineering fees. The officials replied that, although price would not be the only consideration, it would be a very important one. They could not explain to their constituents why the work was awarded to Engineer X if Engineer Y offered to do it for 2 percent less. The civil engineering company declined to participate (Alger, Christensen, and Olmsted, 49).

5 In the following case are Doe's presentation and offer entirely ethical?

John Doe, P.E., a principal in a consulting engineering firm, attended a public meeting of a township board of supervisors which had under consideration a water pollution control project with an estimated construction cost of $7 million. Doe presented a so-called "cost-saving plan" to the supervisors under which his firm would work with the engineering firm retained for the project to find "cost-saving" methods to enable the township to proceed with the project and thereby not lose the federal funding share because of the township's difficulty in financing its share of the project.

Doe further advised the supervisors that his proposal contemplated providing his "cost-saving" services on the basis of being paid ten percent of the savings, but his firm would not be paid any amount if it did not achieve a reduction in the construction cost. Doe added that his firm's value engineering approach would be based on an analysis of the plans and specifications prepared by the design firm and that his operation would not require that the design firm be displaced *(NSPE: Opinions of the Board of Ethical Review,* Case No. 77–10, *Professional Engineer,* Vol. 48, No. 6, June 1978, p. 52).

6 Should Engineer A in the following example take any further action, and if so what type?

During an investigation of a bridge collapse, Engineer A investigates another similar bridge, and finds it to be only marginally safe. He contacts the governmental agency responsible for the bridge and informs them of his concern for the safety of the structure. He is told that the agency is aware of this situation, and has planned to provide in next year's budget for its repair. Until then, the bridge must remain open to traffic. Without this bridge, emergency vehicles such as police and fire apparatus would have to use an alternate route which would increase their response time about twenty minutes.

Engineer A is thanked for his concern and asked to say nothing about the condition of the bridge. The agency is confident that the bridge will be safe (unpublished case study written up by and used with permission of L. R. Smith and Sheri Smith).

7 Examine and assess some recent disputes over whether fee competition among consulting engineers is ethical. A good place to begin is with the articles by Crawford Greene and Gerald Swenson listed in the Bibliography.

SUMMARY

Engineering offers a variety of "existential pleasures": for example, the delights of changing the world, being creative, understanding the physical universe, and promoting the good of others. It also offers economic rewards. Yet in addition there is a direct moral dimension related to the decisions to enter engineering as a profession, to specialize in a particular field of engineering, and to pursue a specific type of work within engineering. How best to meet the demands of one's personal conscience, as well as how best to secure one's personal happiness, should be considered in deciding.

Thus the notion of an ethics of vocation, or a moral philosophy of work, is relevant to career decisions. William Frankena has argued that there is a duty to pursue a vocation. Whether or not that is so, there clearly are special moral duties attached to specific professional roles within engineering. Career decisions involve relating one's basic moral convictions to such professional duties and also to the specific types of activities one will pursue.

Related to an ethics of vocation is the notion of a personal work ethic: that is, the values and ideals according to which one pursues one's career. Another aspect, as applied to engineering, is the assessment of various proposed models for the social role of engineers: for example, savior, guardian, bureaucratic servant, social servant, social enabler and catalyst, and game player.

As examples of moral issues within specific areas of engineering, we surveyed problems in computer technology, industries that pollute the environment, weapons development, and consulting engineering. Another example is nuclear engineering, which was touched on in Chap. 4.

Computer ethics is the branch of engineering ethics dealing with moral issues in computer technology. It is becoming increasingly important as computers become the technological backbone of contemporary society. Many issues arise over the possibility of computer abuse—unethical or illegal conduct in which computers are made to play a central role, as in theft, fraud, and violation of privacy rights. Some issues are merely special examples of those we have dealt with earlier: for example, whistleblowing, safety, informed consent, ownership rights, and conscientious refusal to engage in unethical activity. But still others have to do with the effects of the widespread use of computers in our society: for example, the elimination and transformation of jobs, or the use of computers in surveillance and political campaigns.

Environmental issues relating to industries that pollute pose extreme challenges to identifying, assessing, and dealing with risks. PCB, cadmium, mercury, and asbestos poisoning represent only a few of many tragic instances of environmental pollution. Likewise, the degradation of the Everglades is but one illustration of how natural environments are sometimes treated in a heedlessly destructive manner. All such instances prompt an attitude of cautious experimentation in uncovering technology's negative side effects on the environment.

There are special moral problems intrinsic to the defense industry, such as those related to planned cost overruns, uncritical proliferation of new weapons, and secrecy in military work. Yet those issues appear insignificant compared with the problem of war and peace. And overshadowing all other issues dealt with in this book is the possibility of a nuclear holocaust. Certainly the decision to enter or avoid weapons development as a career is among the most important confronting engineers. It is also one of the most deeply personal decisions one can make and should involve a searching examination of both one's individual conscience and the social and political issues of weapons technology.

Consulting engineers have greater freedom than most salaried engineers. They also must deal with a wider variety of moral concerns since they are responsible for their own business decisions. Honesty in advertising may be difficult to maintain in the face of pressures to promote their business. Controversy still exists concerning the legitimacy of competitive bidding and accepting contingency fees. And special problems arise in determining the extent of responsibility in such areas as making safety inspections and arriving at ways to settle disputes.

CHAPTER 8

CONCLUSION AND FURTHER ISSUES

Before he was elected President of the United States, Herbert Hoover was the engineer-president of the American Institute of Mining Engineers and also of the former Federated American Engineering Societies. Serving in those capacities he had a significant impact on the development of the profession of engineering (Layton, 1971, 179–218). In his memoirs he describes the honors and liabilities of engineering:

> It is a great profession. There is the fascination of watching a figment of the imagination emerge through the aid of science to a plan on paper. Then it moves to realization in stone or metal or energy. Then it brings jobs and homes to men. Then it elevates the standards of living and adds to the comforts of life. That is the engineer's high privilege.
>
> The great liability of the engineer compared to men of other professions is that his works are out in the open where all can see them. His acts, step by step, are in hard substance. He cannot bury his mistakes in the grave like the doctors. He cannot argue them into thin air or blame the judge like the lawyers. He cannot, like the architects, cover his failures with trees and vines. He cannot, like the politicians, screen his shortcomings by blaming his opponents and hope that the people will forget. The engineer simply cannot deny that he did it. If his works do not work, he is damned (Hoover, 132–133).

Hoover is reflecting on a time when engineering was still dominated, at least in outlook, by the independent craftsperson and consultant. When a bridge fell or a ship sank, the specific engineers responsible could be more easily identified. This made it easier to endorse Hoover's vision of individualism in regard both to creativity and personal accountability within engineering.

Today the products of engineering are as much "out in the open" as in Hoover's time. In fact, mass communication assures that mistakes are given even closer public scrutiny. And there are more engineers than ever. Yet despite their greater numbers, the engineers of today are less visible to the public than they were in that earlier era. Technological progress is taken for granted as being the norm, and technological failure is blamed on corporations. And in the public's eye, the representative of any corporation is its top manager, who is often far removed from the daily creative endeavors of the company's engineers.

Today it is thus difficult for engineers to retain a sense of mutual understanding with and accountability to the public. The dominant image of engineers, even for many engineers themselves, has become that of a servant to organizations rather than a public guardian.

We have explored a number of the additional threats to a sense of personal responsibility coming from within organizations. Some of those arise inevitably from the context of contemporary competitive enterprise: fragmentation of work, pressured time schedules, tight budgets. But others are not inevitable: overly rigid lines of authority, disregard of the individual's conscience, suppression of free expression of professional judgment. The latter can be and are being changed as both employee rights and the public responsibilities of corporations receive greater attention.

There is no need to review here the many issues we have explored concerning the responsibilities of engineers. A rereading of the chapter summaries will suffice for that. Yet it is worthwhile to summarize briefly a few of the main themes which have surfaced throughout this book.

SUMMARY OF THEMES

We have emphasized the personal moral autonomy of individuals. We did so with the conviction that responsible conduct is important in all aspects of living, not the least of which is work, and that it cannot be transferred away to employers, code writers, or lawmakers. Yet moral autonomy, as we have treated it, is vastly different from unrestrained individual whim. Autonomous moral agents are sensitive to the legitimate claims of employer authority and the need for reasonable laws and some codified ethical rules. At the same time they are sufficiently concerned to critically assess laws, codes, and directives in the light of reasoned value principles.

We have also emphasized the obligations of engineers to the public based on the public's right to make informed decisions about the risks affecting it. Meeting this obligation requires that engineers working within corporations and other large organizations be given the freedom to express and act on their professional judgment. Thus we were led to stress the importance of employee rights for engineers.

Those rights are limited by the legitimate authority of employers. While we explored several morally relevant aspects of that authority, we offered no simple algorithm for precisely determining its limits. Here, as elsewhere, there is room for disagreement among reasonable people. Engineers would be dogmatic if they always insisted their views were the correct ones. Managers would be equally dogmatic and abusive of their authority if they gave no recognition to the conscientious concern engineers often display and action they sometimes take in disputed areas. There is a need for appeal groups, both within the employing organizations (ombudspersons, ethics committees, etc.) and outside them (confidential review committees of professional societies or the government). Above all there is the need for mutual understanding among engineers and management about the need to cooperatively resolve value conflicts.

These themes can be integrated within a conception of engineering as an experiment on a societal scale involving human subjects. We can restate the points previously made within the framework of this paradigm: As subjects of the "experiment," clients and the public have a right to make informed decisions concerning the risks they will be subject to. As central participants in the "experiment," engineers have a special responsibility to respect those rights. To do so they must be allowed the freedom to conduct the "experiment" in a professional manner, which includes freedom from actions on the part of managers and supervisors that would seriously violate their consciences. On the other hand, managers play an equally central role in engineering "experiments." They have the legitimate authority to guide many areas of the work lives of engineers, although they should avoid undue restrictions on work and (as just noted) refrain from violating the consciences of those they supervise. But it must be kept in mind that moral dilemmas involving competing obligations to employers and the public are virtually inevitable for engineers. Resolving those dilemmas requires both autonomous moral reasoning about and tolerance and sensitivity to different views concerning areas of disagreement that can arise among responsible moral agents.

We have explored a variety of issues in engineering ethics within a framework of philosophical ethical theory. Yet it is important to avoid any suggestion that we have made an exhaustive survey of such issues. The field is far too young and rapidly growing for there to be even a consensus as to what the central topics are. Hence we conclude this book with a sketch of a few of the areas we have either omitted or mentioned only in passing, but which can be gathered under the unifying theme of the profession's place in the life of an engineer. Many of the points we raise in the following will be stated in the form of questions, not because we are reluctant to bare our own views, but because we wish to present the reader with the challenge of applying ethical theory to the issues noted.

RESPONSIBILITIES OF AND TO THE PROFESSION

We have focused on the responsibilities of engineers *in* their work—that is, the responsibilities they have for the projects on which they work and for related aspects of their jobs. But there are important issues pertaining to the responsibilities of the engineering profession as a whole. Many of those issues concern the collective responsibilities of professional and technical societies. There are also related issues concerning the responsibilities of engineers as engineers to their profession and to the world in which they live as citizens.

Collective Responsibilities of the Profession

It makes sense to praise or blame a profession as a whole for either establishing or failing to establish appropriate standards, and for either fostering or discouraging appropriate ideals. It also makes sense to praise or criticize its general role in society. This presupposes that there are general obligations and ideals which a profession as a collective body should pursue.

For one thing, making general appraisals of professions entails examining their ''macro economics'': that is, examining how they do and should function as a group within contemporary society. For example, to what extent is it desirable for the engineering profession as a whole to set standards in such areas as disposal of toxic wastes? (Ladd, 1980, 158)

Or, to take another kind of topic, is the trend toward increasing degrees of professionalism within engineering in the public interest? Here many issues are involved, at least given the model of professionalism derived from developments in medicine and law:

1 Should the engineering profession be allowed to have the authority to decide which students and how many students will be admitted to schools of engineering? Should laypersons representing the public have a say?

2 Should registration and licensing of all engineers be mandatory, as they are for doctors and lawyers? There would be potential benefits: for example, greater assurance that minimal standards of training and skill would be met by all engineers. But there would also be drawbacks, if only that bureaucratic red-tape would increase.

3 Should continuing education be mandatory for all engineers?

These and related issues are given regular, detailed examination at professional conferences. A sample of how they are typically approached can be found in *Ethics, Professionalism, and Maintaining Competence*, edited by Russel C. Jones et al., which is a collection of papers presented at a civil engineering conference. And many professional journals regularly carry articles treating these topics: for example, *Engineering Education*, *Professional Engineer, Engineering Issues (ASCE)*, and *IEEE Spectrum*. A stimulating study of historical and sociological backgrounds for dealing with these topics is *The Credential Society* (Collins). Yet there remain few attempts to approach the issues within a framework of philosophical ethical theory.

Responsibilities of Professional Societies

Many of the large-scale social and economic issues involved in engineering ethics relate directly to professional societies. Professional societies help unify a profession. They also act on behalf of the profession, or at least on behalf of a large segment of the profession. A number of responsibilities ascribed to professions as collectives transfer directly to existing professional societies.

Many of the current tensions in professional societies exist because of uncertainties about their involvement in moral issues. This was illustrated in the BART case, as we saw in Chap. 3. One chapter of the California Society of Professional Engineers felt it should play a role in supporting the efforts of the three engineers who sought to act outside normal organizational channels in serving the public. Another chapter felt it was inappropriate for the society to do so.

It is unlikely that existing professional societies will, and perhaps undesirable that they should, take any univocal proemployee or promanagement stand. Their memberships, after all, are typically a mixture of engineers in management, supervision, and nonmanagement. Yet professional societies can, should, and are increasingly playing a role in conflicts involving moral issues, although rank and file engineers remain skeptical because they still consider the societies management-dominated (Flores, 1982, 80). Through membership participation on committees, they provide a sympathetic and informed forum for hearing opposing viewpoints and making recommendations. Through guidelines they formulate for employment practice and conflict resolution they can help forestall debilitating disputes within corporations. Details concerning the desirable extent and form of such activities deserve on-going discussion within engineering ethics.

Certainly we might expect professional societies increasingly to foster the study of engineering ethics. They are ideal groups for sponsoring ethics workshops, conducting surveys on matters of ethical concern, informing their members of developments related to ethics, and encouraging schools of engineering to support regular and continuing education courses in engineering ethics. Unfortunately a recent study of the activities of professional societies concluded that "little attention and only minimal resources have been directed toward professional ethics matters" (Chalk, Frankel, and Chafer, 101). As at least two other observers have pointed out, progress needs to be made before a genuine community based on a sense of shared values exists in the engineering profession (Perrucci and Gerstl, *Profession without Community: Engineers in American Society*, 176).

Another set of issues concerns the role professional societies should play in political decisions related to engineering and technological development. Should there be lobbying groups at the state and national levels seeking legislation on issues like engineering registration, continuing education requirements, and engineers' rights? Opinions are still sharply divided.

There is also the question of the direction and course of existing and possible new professional organizations. Perhaps it is desirable to have greater unity among engineering societies and thus to encourage newer and higher-level umbrella organizations to arise (such as the recently established American Association of Engineering Societies). Yet there are risks in seeking more unified power and action. Would a single powerful engineering society comparable to the American Medical Association or the American Bar Association be in the public interest?

Ultimately these "macro" issues lead back to the "micro" issues related to individual responsibility. For it is individuals involved in their professional societies who are the ultimate loci of action. An interesting area of engineering ethics would be to examine the typical situations in which individuals might be called upon to perform special services within their professional societies, and how far they are obligated to do so.

This leads us to the related topic of the obligations engineers have to their profession.

Obligations to the Profession

The code of ethics of the former Engineers' Council for Professional Development suggests that engineers should obey the code in order to "uphold and advance the integrity, honor and dignity of the engineering profession." Similarly, the preamble to the code of the National Society of Professional Engineers suggests that the code should be followed in part "to uphold and advance the honor and dignity of the engineering profession." Should such statements be dismissed as remnants of the understandable *esprit de corps* of an emerging major profession? Or are there special professional obligations to the engineering profession which engineers should recognize?

Surely engineers have obligations to their peers, such as not to unfairly criticize or degrade them. They also may have the obligation to participate in technical societies in order to keep themselves up to date in their field and to help assure the continuation of such societies. That obligation derives from another obligation to clients, employers, and the public to advance their own and their colleagues' skills and knowledge. And at the very least it is highly desirable for engineers to be active in the ethics-related activities of their professional societies.

Are there additional obligations to defend the honor of the engineering profession against attacks from its critics? Would not the public and the profession itself be better served through a constant effort to seek truth and fairness in assessments of the engineering profession, and through a willingness to raise trenchant criticisms as conscience dictates? The legitimate honor of any profession as a whole can only be earned through its members acting with integrity in meeting their professional responsibilities.

Surely something can also be said in defense of the idea of a duty to respect and defend the honor of the profession. Effective professional activity, whether in engineering or any other profession, requires a substantial degree of trust from clients and the public. Total absence of such trust would undermine the possibility of making contracts, engaging in cooperative work, exercising professional autonomy free of excessive regulation, and working under humane conditions. Building and sustaining that trust is an important responsibility shared by all engineers.

This does not mean glossing over the genuine problems and faults one may encounter as the member of a profession. Nor does it mean always endorsing the more optimistic view of one's profession. But it does mean taking seriously the fact that appearances and reputation are important and should be given balanced and fair attention. Just what this amounts to in practice for engineers is a legitimate area of study for engineering ethics.

We might add that there is always the danger that the idea of an obligation to one's profession can become perverted into a narrow, self-interested concern. Such would be the case, for example, if a profession deliberately limited the number of its practitioners so as to create a greater demand for them and hence manipulate their salaries or fees upward. For this reason it may be preferable to understand talk about "obligations to the profession" as shorthand for certain obligations to the public. Engineers as individuals and as a group owe it to the public to sustain a professional climate conducive to meeting their other obligations to the public.

Responsibilities of Engineer-Citizens

Do engineers as a group and as individuals have special responsibilities as citizens which go beyond those of nonengineer-citizens? For example, should they participate more actively than others in social debates concerning industrial pollution, automobile safety, and disposal of nuclear waste?

Answering this question would require a clarification of the obligations citizens in general have concerning public policy issues. But even here there is considerable disagreement. One view holds that no one is strictly obligated to participate in public decision making. Instead, such participation is a moral ideal which it is desirable for citizens to embrace and pursue as their time allows. A contrasting view holds that all citizens have an obligation to devote some of their time and energies to public policy matters. Minimal requirements for everyone are to stay informed about issues that can be voted on, while stronger obligations arise for those who by professional background are well grounded in specific issues as well as for those who have the time to train themselves as public advocates.

Engineers are not as well represented on many legislative and advisory bodies as they might be. Perhaps they are too modest about offering their services or maybe they see a number of complications arising from volunteer work.

For engineers in private practice, the latter consideration can be particularly troubling. For example, there is the matter of advertising: While an engineer who is employed by a company will bring recognition and honor to the company through volunteer activities, any such efforts on the part of a self-employed engineer could be interpreted as self-serving attempts to gain publicity and perhaps even to secure valuable inside information.

Another type of problem arises for salaried as well as self-employed engineers who as volunteers are asked to provide detailed technical information or services. At what point should they draw the line and refuse to provide that information or those free services when other engineers could be engaged to do so at regular fees?

Issues arising from making services available for civic purposes are well covered in the book *Ethical Problems in Engineering*, by Philip L. Alger, N. A. Christensen, and Sterling P. Olmsted. Involvement of engineers in charitable and religious organizations is frequently recommended by both professional groups and employers. We ourselves would certainly recommend full participation by engineers in all aspects of community life, but we would warn against the notion that good deeds outside the professional sphere of activity can compensate for unethical conduct inside it.

Study Questions

1 Defend your view as to whether engineers have special obligations beyond those of nonengineers to enter into public debates over technological development.

2 Currently only state, not national, registration is possible for engineers, and even that is frequently voluntary or optional. Assess the following views concerning the desirability of requiring national registration of all engineers.

  a "The method of obtaining national registration as well as the approach would
    act as a vehicle for elevating the engineering profession to standards that
    doctors and lawyers now enjoy" ("Comments from Professional Engineers:
    National Registration," 26).
  b "I feel that a national registration law administered by a Federal agency would
    be very detrimental to the engineering profession because it would transfer our
    local responsibilities and authority to a Federal power. . . . [In] practice I think
    it would become a dictatorial agency, administering the law for its own
    purposes, with very little regard for the engineering profession" ("Comments
    from Professional Engineers: National Registration," 26).
3 Examine several recent issues of professional engineering journals to locate
  articles on ethics. Report on the types of issues being addressed in those articles.

SAMPLE CODES

NSPE
ECPD/ABET
IEEE

NATIONAL SOCIETY OF PROFESSIONAL ENGINEERS

CODE OF ETHICS FOR ENGINEERS

PREAMBLE

Engineering is an important and learned profession. The members of the profession recognize that their work has a direct and vital impact on the quality of life for all people. Accordingly, the services provided by engineers require honesty, impartiality, fairness and equity, and must be dedicated to the protection of the public health, safety and welfare. In the practice of their profession, engineers must perform under a standard of professional behavior which requires adherence to the highest principles of ethical conduct on behalf of the public, clients, employers and the profession.

I. FUNDAMENTAL CANONS

Engineers, in the fulfillment of their professional duties, shall:

1. Hold paramount the safety, health and welfare of the public in the performance of their professional duties.
2. Perform services only in areas of their competence.
3. Issue public statements only in an objective and truthful manner.
4. Act in professional matters for each employer or client as faithful agents or trustees.
5. Avoid improper solicitation of professional employment.

II. RULES OF PRACTICE

1. Engineers shall hold paramount the safety, health and welfare of the public in the performance of their professional duties.

    a. Engineers shall at all times recognize that their primary obligation is to protect the safety, health, property and welfare of the public. If their professional judgment is overruled under circumstances where the safety, health, property or welfare of the public are endangered, they shall notify their employer or client and such other authority as may be appropriate.

    b. Engineers shall approve only those engineering documents which are safe for public health, property and welfare in conformity with accepted standards.

    c. Engineers shall not reveal facts, data or information obtained in a professional capacity without the prior consent of the client or employer except as authorized or required by law or this Code.

d. Engineers shall not permit the use of their name or firm name nor associate in business ventures with any person or firm which they have reason to believe is engaging in fraudulent or dishonest business or professional practices.

e. Engineers having knowledge of any alleged violation of this Code shall cooperate with the proper authorities in furnishing such information or assistance as may be required.

2. Engineers shall perform services only in the areas of their competence.

a. Engineers shall undertake assignments only when qualified by education or experience in the specific technical fields involved.

b. Engineers shall not affix their signatures to any plans or documents dealing with subject matter in which they lack competence, nor to any plan or document not prepared under their direction and control.

c. Engineers may accept an assignment outside of their fields of competence to the extent that their services are restricted to those phases of the project in which they are qualified, and to the extent that they are satisfied that all other phases of such project will be performed by registered or otherwise qualified associates, consultants, or employees, in which case they may then sign the documents for the total project.

3. Engineers shall issue public statements only in an objective and truthful manner.

a. Engineers shall be objective and truthful in professional reports, statements or testimony. They shall include all relevant and pertinent information in such reports, statements or testimony.

b. Engineers may express publicly a professional opinion on technical subjects only when that opinion is founded upon adequate knowledge of the facts and competence in the subject matter.

c. Engineers shall issue no statements, criticisms or arguments on technical matters which are inspired or paid for by interested parties, unless they have prefaced their comments by explicitly identifying the interested parties on whose behalf they are speaking, and by revealing the existence of any interest the engineers may have in the matters.

4. Engineers shall act in professional matters for each employer or client as faithful agents or trustees.

a. Engineers shall disclose all known or potential conflicts of interest to their employers or clients by promptly informing them of any business association, interest, or other circumstances which could influence or appear to influence their judgment or the quality of their services.

b. Engineers shall not accept compensation, financial or otherwise, from more than one party for services on the same project, or for services pertaining to the same project, unless the circumstances are fully disclosed to, and agreed to, by all interested parties.

c. Engineers shall not solicit or accept financial or other valuable consideration, directly or indirectly, from contractors, their agents, or other parties in connection with work for employers or clients for which they are responsible.

d. Engineers in public service as members, advisors or employees of a governmental body or department shall not participate in decisions with respect to professional services solicited or provided by them or their organizations in private or public engineering practice.

e. Engineers shall not solicit or accept a professional contract from a governmental body on which a principal or officer of their organization serves as a member.

5. Engineers shall avoid improper solicitation of professional employment.

   a. Engineers shall not falsify or permit misrepresentation of their, or their associates', academic or professional qualifications. They shall not misrepresent or exaggerate their degree of responsibility in or for the subject matter of prior assignments. Brochures or other presentations incident to the solicitation of employment shall not misrepresent pertinent facts concerning employers, employees, associates, joint venturers or past accomplishments with the intent and purpose of enhancing their qualifications and their work.

   b. Engineers shall not offer, give, solicit or receive, either directly or indirectly, any political contribution in an amount intended to influence the award of a contract by public authority, or which may be reasonably construed by the public of having the effect or intent to influence the award of a contract. They shall not offer any gift, or other valuable consideration in order to secure work. They shall not pay a commission, percentage or brokerage fee in order to secure work except to a bona fide employee or bona fide established commercial or marketing agencies retained by them.

III. PROFESSIONAL OBLIGATIONS

1. Engineers shall be guided in all their professional relations by the highest standards of integrity.

   a. Engineers shall admit and accept their own errors when proven wrong and refrain from distorting or altering the facts in an attempt to justify their decisions.

   b. Engineers shall advise their clients or employers when they believe a project will not be successful.

   c. Engineers shall not accept outside employment to the detriment of their regular work or interest. Before accepting any outside employment they will notify their employers.

   d. Engineers shall not attempt to attract an engineer from another employer by false or misleading pretenses.

   e. Engineers shall not actively participate in strikes, picket lines, or other collective coercive action.

   f. Engineers shall avoid any act tending to promote their own interest at the expense of the dignity and integrity of the profession.

2. Engineers shall at all times strive to serve the public interest.

   a. Engineers shall seek opportunities to be of constructive service in civic affairs and work for the advancement of the safety, health and well-being of their community.

   b. Engineers shall not complete, sign, or seal plans and/or specifications that are not of a design safe to the public health and welfare and in conformity with accepted engineering standards. If the client or employer insists on such unprofessional conduct, they shall notify the proper authorities and withdraw from further service on the project.

   c. Engineers shall endeavor to extend public knowledge and appreciation of engineering and its achievements and to protect the engineering profession from misrepresentation and misunderstanding.

3. Engineers shall avoid all conduct or practice which is likely to discredit the profession or deceive the public.

   a. Engineers shall avoid the use of statements containing a material misrepresentation of fact or omitting a material fact necessary to keep statements from being misleading; statements intended or likely to create an unjustified expectation; statements containing prediction of future success; statements containing an opinion as to the quality of the Engineers' services; or statements intended or likely to attract clients by the use of showmanship, puffery, or self-laudation, including the use of slogans, jingles, or sensational language or format.

b. Consistent with the foregoing, Engineers may advertise for recruitment of personnel.

c. Consistent with the foregoing, Engineers may prepare articles for the lay or technical press, but such articles shall not imply credit to the author for work performed by others.

4. Engineers shall not disclose confidential information concerning the business affairs or technical processes of any present or former client or employer without his consent.

a. Engineers in the employ of others shall not without the consent of all interested parties enter promotional efforts or negotiations for work or make arrangements for other employment as a principal or to practice in connection with a specific project for which the Engineer has gained particular and specialized knowledge

b. Engineers shall not, without the consent of all interested parties, participate in or represent an adversary interest in connection with a specific project or proceeding in which the Engineer has gained particular specialized knowledge on behalf of a former client or employer.

5. Engineers shall not be influenced in their professional duties by conflicting interests

a. Engineers shall not accept financial or other considerations, including free engineering designs, from material or equipment suppliers for specifying their product.

b. Engineers shall not accept commissions or allowances, directly or indirectly, from contractors or other parties dealing with clients or employers of the Engineer in connection with work for which the Engineer is responsible.

6. Engineers shall uphold the principle of appropriate and adequate compensation for those engaged in engineering work.

a. Engineers shall not accept remuneration from either an employee or employment agency for giving employment.

b. Engineers, when employing other engineers, shall offer a salary according to professional qualifications and the recognized standards in the particular geographical area.

7. Engineers shall not compete unfairly with other engineers by attempting to obtain employment or advancement or professional engagements by taking advantage of a salaried position, by criticizing other engineers, or by other improper or questionable methods.

a. Engineers shall not request, propose, or accept a professional commission on a contingent basis under circumstances in which their professional judgment may be compromised.

b. Engineers in salaried positions shall accept part-time engineering work only at salaries not less than that recognized as standard in the area.

c. Engineers shall not use equipment, supplies, laboratory, or office facilities of an employer to carry on outside private practice without consent.

8. Engineers shall not attempt to injure, maliciously or falsely, directly or indirectly, the professional reputation, prospecs, practice or employment of other engineers, nor indiscriminately criticize other engineers' work. Engineers who believe others are guilty of unethical or illegal practice shall present such information to the proper authority for action.

a. Engineers in private practice shall not review the work of another engineer for the same client, except with the knowledge of such engineer, or unless the connection of such engineer with the work has been terminated.

b. Engineers in governmental, industrial or educational employ are entitled to review and evaluate the work of other engineers when so required by their employment duties.

c. Engineers in sales or industrial employ are entitled to make engineering comparisons of represented products with products of other suppliers.

9. Engineers shall accept personal responsibility for all professional activities.

a. Engineers shall conform with state registration laws in the practice of engineering.

b. Engineers shall not use association with a nonengineer, a corporation, or partnership, as a "cloak" for unethical acts, but must accept personal responsibility for all professional acts.

10. Engineers shall give credit for engineering work to those to whom credit is due, and will recognize the proprietary interests of others.

a. Engineers shall, whenever possible, name the person or persons who may be individually responsible for designs, inventions, writings, or other accomplishments.

b. Engineers using designs supplied by a client recognize that the designs remain the property of the client and may not be duplicated by the Engineer for others without express permission.

c. Engineers, before undertaking work for others in connection with which the Engineer may make improvements, plans, designs, inventions, or other records which may justify copyrights or patents, should enter into a positive agreement regarding ownership.

d. Engineers' designs, data, records, and notes referring exclusively to an employer's work are the employer's property.

11. Engineers shall cooperate in extending the effectiveness of the profession by interchanging information and experience with other engineers and students, and will endeavor to provide opportunity for the professional development and advancement of engineers under their supervision.

a. Engineers shall encourage engineering employees' efforts to improve their education.

b. Engineers shall encourage engineering employees to attend and present papers at professional and technical society meetings.

c. Engineers shall urge engineering employees to become registered at the earliest possible date.

d. Engineers shall assign a professional engineer duties of a nature to utilize full training and experience, insofar as possible, and delegate lesser functions to subprofessionals or to technicians.

e. Engineers shall provide a prospective engineering employee with complete information on working conditions and proposed status of employment, and after employment will keep employees informed of any changes.

"By order of the United States District Court for the District of Columbia, former Section 11(c) of the NSPE Code of Ethics prohibiting competitive bidding, and all policy statements, opinions, rulings or other guidelines interpreting its scope, have been rescinded as unlawfully interfering with the legal right of engineers, protected under the antitrust laws, to provide price information to prospective clients; accordingly, nothing contained in the NSPE Code of Ethics, policy statements, opinions, rulings or other guidelines prohibits the submission of price quotations or competitive bids for engineering services at any time or in any amount."

Statement by NSPE Executive Committee

In order to correct misunderstandings which have been indicated in some instances since the issuance of the Supreme Court decision and the entry of the Final Judgment, it is noted that in its decision of April 25, 1978, the Supreme Court of the United States declared: "The Sherman Act does not require competitive bidding."

It is further noted that as made clear in the Supreme Court decision:

1. Engineers and firms may individually refuse to bid for engineering services.
2. Clients are not required to seek bids for engineering services.
3. Federal, state, and local laws governing procedures to procure engineering services are not affected, and remain in full force and effect.
4. State societies and local chapters are free to actively and aggressively seek legislation for professional selection and negotiation procedures by public agencies.
5. State registration board rules of professional conduct, including rules prohibiting competitive bidding for engineering services, are not affected and remain in full force and effect. State registration boards with authority to adopt rules of professional conduct may adopt rules governing procedures to obtain engineering services.
6. As noted by the Supreme Court, "nothing in the judgment prevents NSPE and its members from attempting to influence governmental action. . . ."

Note:
In regard to the question of application of the Code to corporations vis-a-vis real persons, business form or type should not negate nor influence conformance of individuals to the Code. The Code deals with professional services, which services must be performed by real persons. Real persons in turn establish and implement policies within business structures. The Code is clearly written to apply to the Engineer and it is incumbent on a member of NSPE to endeavor to live up to its provisions. This applies to all pertinent sections of the Code.

NSPE Publication No. 1102 As revised, July 1981

[Used with permission of N.S.P.E.]

ENGINEERS' COUNCIL FOR PROFESSIONAL DEVELOPMENT*

CODE OF ETHICS OF ENGINEERS

THE FUNDAMENTAL PRINCIPLES

Engineers uphold and advance the integrity, honor and dignity of the engineering profession by:

I.   using their knowledge and skill for the enhancement of human welfare;

II. being honest and impartial, and serving with fidelity the public, their employers and clients;

III. striving to increase the competence and prestige of the engineering profession; and

IV. supporting the professional and technical societies of their disciplines.

THE FUNDAMENTAL CANONS

1. Engineers shall hold paramount the safety, health and welfare of the public in the performance of their professional duties.

2. Engineers shall perform services only in the areas of their competence.

3. Engineers shall issue public statements only in an objective and truthful manner.

4. Engineers shall act in professional matters for each employer or client as faithful agents or trustees, and shall avoid conflicts of interest.

5. Engineers shall build their professional reputation on the merit of their services and shall not compete unfairly with others.

6. Engineers shall act in such a manner as to uphold and enhance the honor, integrity and dignity of the profession.

7. Engineers shall continue their professional development throughout their careers and shall provide opportunities for the professional development of those engineers under their supervision.

*Now the Accreditation Board for Engineering and Technology (ABET), which has recommended that the American Association of Engineering Societies adopt this code. The latter is considering modifications.

INSTITUTE OF ELECTRICAL AND ELECTRONICS ENGINEERS

PREAMBLE

Engineers, scientists and technologists affect the quality of life for all people in our complex technological society. In the pursuit of their profession, therefore, it is vital that IEEE members conduct their work in an ethical manner so that they merit the confidence of colleagues, employers, clients and the public. This IEEE Code of Ethics represents such a standard of professional conduct for IEEE members in the discharge of their responsibilities to employers, to clients, to the community and to their colleagues in this Institute and other professional societies.

ARTICLE I

Members shall maintain high standards of diligence, creativity and productivity, and shall:

1 Accept responsibility for their actions;

2 Be honest and realistic in stating claims or estimates from available data;

3 Undertake technological tasks and accept responsibility only if qualified by training or experience, or after full disclosure to their employers or clients of pertinent qualifications;

4 Maintain their professional skills at the level of the state of the art, and recognize the importance of current events in their work;

5 Advance the integrity and prestige of the profession by practicing in a dignified manner and for adequate compensation.

ARTICLE II

Members shall, in their work:

1 Treat fairly all colleagues and co-workers, regardless of race, religion, sex, age or national origin;

2 Report, publish and disseminate freely information to others, subject to legal and proprietary restraints;

3 Encourage colleagues and co-workers to act in accord with this Code and support them when they do so;

4 Seek, accept and offer honest criticism of work, and properly credit the contributions of others;

5 Support and participate in the activities of their professional societies;

6 Assist colleagues and co-workers in their professional development.

ARTICLE III

Members shall, in their relations with employers and clients;

1 Act as faithful agents or trustees for their employers or clients in professional and business matters, provided such actions conform with other parts of this Code;

2 Keep information on the business affairs or technical processes of an employer or client in confidence while employed, and later, until such information is properly released, provided such actions conform with other parts of this Code;

3 Inform their employers, clients, professional societies or public agencies or private agencies of which they are members or to which they may make presentations, of any circumstance that could lead to a conflict of interest;

4 Neither give nor accept, directly or indirectly, any gift, payment or service of more than nominal value to or from those having business relationships with their employers or clients;

5 Assist and advise their employers or clients in anticipating the possible consequences, direct and indirect, immediate or remote, of the projects, work or plans of which they have knowledge.

ARTICLE IV

Members shall, in fulfilling their responsibilities to the community:

1 Protect the safety, health and welfare of the public and speak out against abuses in these areas affecting the public interest;

2 Contribute professional advice, as appropriate, to civic, charitable or other nonprofit organizations;

3 Seek to extend public knowledge and appreciation of the profession and its achievements.

Approved February 18, 1979, by the Board of Directors of the Institute of Electrical and Electronics Engineers, Inc.

BIBLIOGRAPHY

Alderman, Frank E., and Robert A. Schultz: *Ethical Problems in Consulting Engineering* (unpublished manuscript). Quotations in text used with permission of authors.

Alger, Philip L., N.A. Christensen, and Sterling P. Olmsted: *Ethical Problems in Engineering*, Wiley, New York, 1965. Quotations used with permission of publisher.

Anderson, Robert M., Robert Perrucci, Dan E. Schendel, and Leon E. Trachtman: *Divided Loyalties*, Purdue, West Lafayette, Indiana, 1980.

Aristotle: *The Nicomachean Ethics*, in Richard McKeon (ed.), *The Basic Works of Aristotle*, Random House, New York, 1941.

Arnould, Richard J., and Henry Grabowski: "Auto Safety Regulation: An Analysis of Market Failure," in *The Bell Journal of Economics*, Vol. 12, No. 1, Spring 1981, pp. 27–48.

Backhouse, Constance: *Sexual Harassment on the Job*, Prentice-Hall, Englewood Cliffs, New Jersey, 1981.

Baier, Kurt, and Nicholas Rescher (eds.): *Values and the Future: The Impact of Technological Change on American Values*, Free Press, New York, 1969.

Bailey, Martin J.: *Reducing Risks to Life: Measurement of the Benefits*, American Enterprise Institute for Public Policy Research, Washington, D.C., 1980

Bailyn, Lotte: *Living with Technology: Issues at Mid-Career*, MIT, Cambridge, Massachusetts, 1980. Quotation in text used with permission of publisher.

Bane, Charles A.: *The Electrical Equipment Conspiracy*, Federal Legal Publications, New York, 1973.

Baram, Michael S.: "Trade Secrets: What Price Loyalty?" *Harvard Business Review*, November-December, 1968; reprinted in Barry, op. cit. (see below), pp. 207–216.

Barnard, Chester I.: *The Functions of the Executive,* Harvard, Cambridge, Massachusetts, 1968.

Barnett, Chris: "Blowing the Whistle on Nuclear Power: Three Leave GE," *New Engineer,* May 1976, pp. 34–42.

Barry, Vincent: *Moral Issues in Business,* Wadsworth, Belmont, California, 1979.

Bartlett, Robert V.: *The Reserve Mining Controversy,* Indiana University Press, Bloomington, 1980.

Barzun, Jacques: "The Professions Under Siege," *Harpers,* Vol. 257, No. 1541, October 1978, pp. 61–67.

Baum, Robert J.: *Ethics and Engineering Curricula,* The Hastings Center, New York, 1980.

Baum, Robert J. (ed.): *Ethical Problems in Engineering,* 2d ed., Volume Two: *Cases,* Center for the Study of the Human Dimensions of Science and Technology, Rensselaer Polytechnic University, Troy, New York 1980.

Baum, Robert J.: "The Limits of Professional Responsibility," in D.L. Babcock and C.A. Smith (eds.), *Values and The Public Works Professional,* 1980, available from the American Public Works Association, 1313 E. 60th St., Chicago, Illinois 60637.

Bayles, Michael D.: *Professional Ethics,* Wadsworth, Belmont, California, 1981.

Bazelon, David L.: "Risk and Responsibility," *Science,* Vol. 205, July 20, 1979, pp. 277–280. Quotation used with permission of author.

Beauchamp, Tom L., and Norman E. Bowie (eds.): *Ethical Theory and Business,* Prentice-Hall, Englewood Cliffs, New Jersey, 1979.

Beauchamp, Tom L.: "The Justification of Reverse Discrimination in Hiring," in ibid., pp. 625–635.

Benn, Stanley I.: "Privacy, Freedom, and Respect for Persons," in J. Roland Pennock and John W. Chapman (eds.), *Nomos XIII,* Atherton, New York, 1971, pp. 1–26.

Blackstone, William T.: "On Rights and Responsibilities Pertaining to Toxic Substances and Trade Secrecy," *The Southern Journal of Philosophy,* Vol. 16, 1978, pp. 589–603.

Blades, Lawrence E.: "Employment at Will Vs. Individual Freedom: On Limiting the Abusive Exercise of Employer Power," *Columbia Law Review,* Vol. 67, 1967, pp. 1404–1435.

Blumberg, Phillip I.: "Corporate Responsibility and the Employee's Duty of Loyalty and Obedience: A Preliminary Inquiry," *Oklahoma Law Review,* Vol. 24, August 1971, pp. 279–318. Condensed version printed in Beauchamp and Bowie, op. cit., pp. 304–317.

Boffey, Philip M.: "Teton Dam Verdict: Foul-up by the Engineers," *Science,* Vol. 195, Jan. 21, 1977, pp. 270–272.

Bok, Sissela: "Whistleblowing and Professional Responsibilities," in Daniel Callahan and Sissela Bok (eds.), *Ethics Teaching in Higher Education,* Plenum, New York, 1980, pp. 277–295.

Borrelli, Peter, Mahlon Easterling, Burton H. Klein, Lester Lees, Guy Pauker, and Robert Poppe: *People, Power and Pollution: Environmental and Public Interest Aspects of Electric Power Plant Siting,* Environmental Quality Lab, California Institute of Technology, Pasadena, 1971.

Boulle, Pierre: *The Bridge Over the River Kwai,* trans. Xan Fielding, Vanguard, New York, 1954.

Bowie, Norman: *Business Ethics,* Prentice-Hall, Englewood Cliffs, New Jersey, 1982.

Brandt, Richard B.: *Ethical Theory,* Prentice-Hall, Englewood Cliffs, New Jersey, 1959.

Buehler, John P.: "Allocation of Risks and Resolution of Disputes Under the Contract and Beyond the Contract," in *Exploratory Study on Responsibility Liability, and Accountability for Risks in Construction,* National Research Council, National Academy of Sciences, Washington, D.C., 1978, pp. 91–100.

Burdick, Eugene, and Harvey Wheeler: *Fail-Safe,* Dell, New York, 1962.

Burke, John G.: "Bursting Boilers and the Federal Power," in M. Kranzberg and W.H. Davenport (eds.), *Technology and Culture,* Schocken, New York, pp. 93–118.

Burton, John F.: "Public Sector Strikes," in Richard T. De George and Joseph A. Pichler (eds.), *Ethics, Free Enterprise and Public Policy,* Oxford, New York, 1978, pp. 127–154.

Callaghan, Dennis W., and Arthur Elkins (eds.): *A Managerial Odyssey: Problems in Business and Its Environment,* 2d ed., Addison-Wesley, Reading, Massachusetts, 1978.

Camps, Frank: "Warning an Auto Company about an Unsafe Design," in Alan F. Westin, *Whistle-Blowing!* (see below), pp. 119–129.

Carter, Charles M.: "Trade Secrets and the Technical Man," *IEEE Spectrum,* Vol. 6, No. 2, February 1969, pp. 51–55.

Chalk, Rosemary, Mark S. Frankel, and Sallie B. Chafer (eds.): *AAAS Professional Ethics Project: Professional Ethics Activities in the Scientific and Engineering Societies,* American Association for the Advancement of Science, Washington, D.C., 1980.

Cherrington, David J.: *The Work Ethic,* AMACOM, New York, 1980.

Cogan, Morris L.: "The Problem of Defining a Profession," *Annals of the American Academy of Political Science,* Vol. 297, 1955, pp. 105–111.

Cohen, Carl: "Why Racial Preference is Illegal and Immoral," in Burton M. Leiser (ed.), *Values in Conflict,* Macmillan, New York, 1981, pp. 373–395.

Cohen, Richard M., and Jules Witcover: *A Heartbeat Away: The Investigation and Resignation of Vice President Spiro T. Agnew,* Viking, New York, 1974.

Collins, Randall: *The Credential Society,* Academic, New York, 1979.

"Comments from Professional Engineers: National Registration," *The American Engineer,* July 1964, pp. 25–30.

Comparato, Frank E.: *Age of Great Guns,* Stackpole, Harrisburg, Pennsylvania, 1965.

Conference on Engineering Ethics, American Society of Civil Engineers, New York, 1975.

Congressional Record: "Nuclear Regulatory Commission's Safety and Licensing Procedures," U.S. Senate Committee on Government Operations, December 13, 1976. Printed in Baum, *Ethical Problems in Engineering,* op. cit., pp. 92–102.

Cooper, B. S., and D. P. Rice: "The Economic Cost of Illness Revisited," *Social Security Bulletin,* Vol. 39, February 1976, pp. 21–36.

Council for Science and Society: *The Acceptability of Risks,* Barry Rose Publishers Ltd., Ringwood, Hants, England, 1977.

Davies, Sandra L.: "Does Sex Make a Difference in Engineering Careers?" *Professional Engineer,* Vol. 51, March 1981, pp. 30–32.

Da Vinci, Leonardo: *The Notebooks of Leonardo Da Vinci,* Vol. I, Edward MacCurdy (ed.), George Braziller, New York, 1939.

Davis, Ruth M.: "Preventative Technology: A Cure for Specific Ills," *Science,* Vol. 188, April 18, 1975. Quotation in text used with permission of author.

De Camp, L. Sprague: *The Ancient Engineers,* Doubleday, Garden City, New York, 1963.

De George, Richard T.: "Ethical Responsibilities of Engineers in Large Organizations: The Pinto Case," *Business and Professional Ethics Journal,* Vol. 1, No. 1, Fall 1981, pp. 1–14.

De Havilland, Sir G., and P. B. Walker, "The Comet Failure," adapted by R.R. Whyte for *Engineering Through Trouble,* The Institution of Mechanical Engineers, London, 1975. Quotations in text used with permission of the Council of the Institution of Mechanical Engineers.

DeVine, John C.: "A Progress Report: Cleaning Up TMI," *IEEE Spectrum,* Vol. 18, No. 3, March 1981, pp. 44–49.

Dilley, Dean M.: "The Anxieties of Vulnerability: Liability Protection for Engineers in the Public Sector," *Professional Engineer,* Vol. 49, No. 5, May 1979, pp. 12–14.

Donaldson, Thomas: *Corporations and Morality,* Prentice-Hall, Englewood Cliffs, New Jersey, 1982.

Donaldson, Thomas, and Patricia H. Werhane (eds.): *Ethical Issues in Business,* Prentice-Hall, Englewood Cliffs, New Jersey, 1979.

Drucker, Peter F.: *Management,* Harper and Row, New York, 1973.

Ducasse, C. J.: *A Philosophical Scrutiny of Religion,* Ronald, New York, 1953. Quotation in text used with permission of publisher.

Durbin, Paul T. (ed.): *A Guide to the Culture of Science, Technology, and Medicine,* Free Press, New York, 1980.

Dworkin, Ronald: *Taking Rights Seriously,* Harvard, Cambridge, Massachusetts, 1977.

Dynes, R.R.: *Organized Behavior in Disaster,* Heath, Lexington, Massachusetts, 1970.

Eddy, Paul, Elaine Potter, and Bruce Page: *Destination Disaster: From the Tri-Motor to the DC-10, The Risk of Flying,* Quadrangle, New York, 1976.

"Edgerton Case," Reports of the IEEE-CSIT Working Group on Ethics & Employment and the IEEE Member Conduct Committee in the matter of Virginia Edgerton's dismissal as information scientist of New York City. Reproduced in *Technology and Society,* No. 22, June 1978, pp. 3–10. See also *The Institute,* news supplement to *IEEE Spectrum,* June 1979, p. 6, for articles on her IEEE Award for Outstanding Public Service.

Elliston, Frederick A.: "Anonymous Whistleblowing," *Business and Professional Ethics Journal,* Vol. 1, No. 2, 1982, pp. 39–58.

Elliston, Frederick A.: "Civil Disobedience and Whistleblowing," *Journal of Business Ethics,* Vol. 1, No. 1, 1982, pp. 23–28.

Engelbrecht, H.C., and F.C. Hanighen: *Merchants of Death: A Study of the International Armament Industry,* Dodd, Mead, New York, 1934.

Ewing, David W.: *Freedom Inside the Organization,* McGraw-Hill, New York, 1977. Quotation in text used with permission of author and publisher.

Ewing, David W.: "IBM's Guidelines to Employee Privacy," *Harvard Business Review,* Vol. 54, 1976, pp. 82–90.

Ewing, David W.: "What Business Thinks about Employee Rights," *Harvard Business Review,* Vol. 55, 1977; reprinted in Westin and Salisbury op. cit. (see below), pp. 21–42. Quotation in text used with permission of author.

Fairweather, Virginia: "$80,000 in Payoffs: An Engineer Tells His Story," *Civil Engineering,* Vol. 48, No. 1, January 1978, pp. 54–55.

Faulkner, Peter: "Exposing Risks of Nuclear Disaster," in Alan F. Westin, *Whistle-Blowing!* (see below), pp. 39–54.

Feldman, Fred: *Introductory Ethics,* Prentice-Hall, Englewood Cliffs, New Jersey, 1978.

Firmage, D. Allan: *Modern Engineering Practice,* Garland STPM Press, New York, 1980.

Fitzgerald, A. Ernest: *The High Priests of Waste,* Norton, New York, 1972.

Fitzgerald, Donald: "The Life and Times of Lawrence Tate," in Baum, *Ethical Problems in Engineering,* Volume Two, pp. 197–198.

Flores, Albert and Deborah G. Johnson: "Collective Responsibility and Professional Roles," *Ethics* (forthcoming).

Flores, Albert (ed.): *Designing for Safety: Engineering Ethics in Organizational Contexts,* Workshop on Engineering Ethics: Designing for Safety, Rensselaer Polytechnic Institute, Troy, New York, 1982.

Flores, Albert: "Engineering Ethics in Organizational Contexts: A Case Study—National Aeronautics and Space Administration," in ibid., pp. 41–82.

Flores, Albert (ed.): *Ethical Problems in Engineering,* 2d ed., Volume One: *Readings,* Center for the Study of the Human Dimensions of Science and Technology, Rensselaer Polytechnic Institute, Troy, 1980.

Flores, Albert: "Engineers' Professional Rights," in ibid., pp. 171–176.

Florman, Samuel C.: *The Existential Pleasures of Engineering,* St. Martin's, New York, 1976.

Florman, Samuel C.: "Moral Blueprints," *Harpers,* Vol. 257, No. 1541, October 1978, pp. 30–33.

Ford, Daniel: "Three Mile Island, 1—Class Nine Accident," *The New Yorker,* Vol. 57, No. 7, April 6, 1981, pp. 49–120.

Ford, Daniel: "Three Mile Island, 2—The Paper Trail," *The New Yorker,* Vol. 57, No. 8, April 13, 1981, pp. 46–109.

Frankena, William K.: *Ethics,* 2d ed., Prentice-Hall, Englewood Cliffs, New Jersey, 1973.

Frankena, William K.: "The Philosophy of Vocation," *Thought,* Vol. 51, December 1976, pp. 393–408.

Freedman, Benjamin: "A Meta-Ethics for Professional Morality," *Ethics,* Vol. 89, 1978, pp. 1–19.

Freedman, Benjamin: "What Really Makes Professional Morality Different: Response to Martin," *Ethics,* Vol. 91, 1981, pp. 626–630.

French, Peter: "What is Hamlet to McDonnell-Douglas or McDonnell-Douglas to Hamlet: DC-10," *Business and Professional Ethics Journal,* Vol. 1, No. 2, 1982, pp. 1–13.

Fried, Charles: *An Anatomy of Values,* Harvard, Cambridge, Massachusetts, 1970.

Friedlander, Gordon: "Bigger Bugs in BART?" *IEEE Spectrum,* Vol. 10, No. 3, March 1973, pp. 32–37.

Friedlander, Gordon: "A Prescription for BART," *IEEE Spectrum,* Vol. 10, No. 4, April 1973, pp. 40–44.

Friedlander, Gordon: "Nuclear Power Plant Safety," *IEEE Spectrum,* Vol. 13, No. 5, May 1976, pp. 70–75.

Friedman, Milton: *Capitalism and Freedom,* The University of Chicago Press, Chicago, 1962.

Friedman, Milton: "The Social Responsibility of Business Is to Increase Its Profits," *New York Times Magazine,* September 13, 1970. Reprinted in Donaldson and Werhane, op. cit., pp. 191–197.

Fruchtbaum, Harold: "Engineers and the Commonweal: Notes Toward a Reformation," in Flores, *Ethical Problems in Engineering,* Vol. One, pp. 257–259.

Fuller, John Grant: *The Gentlemen Conspirators,* Grove Press, New York, 1962.

Galbraith, John Kenneth: *The New Industrial State,* revised 2d ed., New American Library, New York, 1971.

Gansler, J.: *The Defense Industry,* MIT, Cambridge, Massachusetts, 1980.

Garrett, Thomas M. et al.: *Cases in Business Ethics,* Appleton Century Crofts, New York, 1968.

Geis, Gilbert: "The Heavy Electrical Equipment Antitrust Cases of 1961," in Gilbert Geis and Robert F. Meier (eds.), *White-Collar Crime: Offenses in Business, Politics, and the Professions,* revised edition, Free Press, New York, 1977, pp. 117–132.

Golding, William: *The Spire,* Harcourt, Brace & World, New York, 1964.

Goldman, Alan H.: *Justice and Reverse Discrimination,* Princeton University Press, Princeton, New Jersey, 1979.

Goldman, Alan H.: *The Moral Foundations of Professional Ethics,* Rowman and Littlefield, New Jersey, 1980.

Graham, F., Jr.: *Since Silent Spring,* Houghton Mifflin, Boston, 1970.

Gray, Mike, and Ira Rosen: *The Warning: Accident at Three Mile Island,* New York, W. W. Norton and Co., 1982.

Greene, Crawford, Jr.: "The Case Against Fee Competition," *Consulting Engineer,* Vol. 52, January 1979.

Greene, Graham: *A Burnt-Out Case,* Penguin, New York, 1977.

Grundy, Richard, Hanno C. Weisbrod, and Samuel S. Epstein: "Toxic Substances," Chap. 3 in *Consumer Health and Product Hazards—Chemicals, Electronic Products, Radiation,* Vol. I of S.S. Epstein and R.D. Grundy (eds.), *The Legislation of Product Safety,* MIT, Cambridge, 1974, pp. 102–169.

Hammer, W.: *Product Safety Management and Engineering,* Prentice-Hall, Englewood Cliffs, New Jersey, 1980. Diagrams adapted with permission of publisher.

Hammurabi: *The Code of Hammurabi,* trans. R. F. Harper, Chicago, University of Chicago Press, 1904, pp. 80–83.

Hardin, Garrett: *Exploring New Ethics for Survival,* Viking, New York, 1968. Quotations in text used with permission of author.

Harding, C. Francis, and Donald T. Canfield: *Legal and Ethical Phases of Engineering,* McGraw-Hill, New York, 1936.

Harris, Robert C., Christoph Hohenemser, and Robert W. Kates: "The Burden of Technological Hazards," in G.T. Goodman and W.D. Rowe (eds.), *Energy Risk Management,* Academic, New York, 1979.

Hart, H.L.A.: *Punishment and Responsibility,* Clarendon, Oxford, 1973.

Haugen, Edward B.: *Probabilistic Approaches to Design,* Wiley, New York, 1968.

Haydon, Graham: "On Being Responsible," *The Philosophical Quarterly,* Vol. 28, 1978, pp. 46–57.

Hayward, David: "Lone Engineer Awaits Verdict on Florida Condominium Collapse," *NCE International,* May 1981, pp. 36–40.

Hehir, J. Bryan: "The Relationship of Moral and Strategic Arguments in the Defense Debate," in Paul T. Durbin (ed.), *Research in Philosophy and Technology,* Vol. 3, JAI Press, Greenwich, Connecticut, 1980, pp. 367–383.

Heilbroner, Robert L., et al.: *In the Name of Profit,* Doubleday, Garden City, 1972.

Herling, John: *The Great Price Conspiracy,* Robert B. Luce, Washington, D.C., 1962.

Hiltzig, Michael A.: "Case Gives Rare Glimpse of Silicon Valley Intrigue," *Los Angeles Times,* Part I, p. 1, March 22, 1982.

Hoover, Herbert: "The Profession of Engineering," *The Memoirs of Herbert Hoover,* Vol. 1, Macmillan, New York, 1961, pp. 131–134. Quotation in text used with permission of the Hoover Foundation.

Houston, Carl W.: "Experiences of a Responsible Engineer," *Conference on Engineering Ethics,* American Society of Civil Engineers, New York, 1975, pp. 25–30.

Howard, Robert T.: "A Bill of Professional Rights for Employed Engineers?" *The American Engineer,* Vol. 36, No. 10, October 1966, pp. 47–50.

Hughson, Roy V. and Philip M. Kohn: "Ethics," *Chemical Engineering,* Vol. 87, No. 19, September 22, 1980, pp. 132–147. Quotations in text used with permission of McGraw-Hill Book Co.

Ibsen, Henrik: *The Master Builder and Other Plays,* trans. Una Ellis-Fermor, Penguin, New York, 1973.

Iijima Nobuko: *Pollution Japan,* Pergamon, New York, 1979.

James, Gene G.: "Whistle Blowing: Its Nature and Justification," *Philosophy in Context,* Vol. 10, 1980, pp. 99–117.

Jansen, Robert B.: *Dams and Public Safety,* Water and Power Resources Service, U.S. Department of the Interior, Denver, 1980.

Jenkins, A. H.: *Adam Smith Today,* Kennikat, Port Washington, New York, 1948.

Jones, Russel C. et al. (eds.): *Ethics, Professionalism and Maintaining Competence,* American Society of Civil Engineers, New York, 1977.

Kahn, Herman, William Brown, and Leon Martel: *The Next 200 Years,* Morrow, New York, 1976.

Kant, Immanuel: *Foundations of the Metaphysics of Morals, with Critical Essays,* Robert Paul Wolff (ed.), Bobbs-Merrill, 1969.

Kaplan, Gadi and Ronald K. Jurgen: "Nuclear Power Plant Safety—I," *IEEE Spectrum,* Vol. 13, No. 5, May 1976, pp. 52–69.

Kardos, Geza: *Heron Road Bridge,* Parts A-B-C-D, Engineering Case Library, ECL 133, Stanford University, Stanford, California, 1969.

Kates, Robert W. (ed.): *Managing Technological Hazard: Research Needs and Opportunities,* Institute of Behavioral Science, University of Colorado, Boulder, 1977.

Keisling, Bill: *Three Mile Island: Turning Point,* Veritas Books, Inc., Seattle, Washington, 1980.

Kelly, Arthur L.: "Italian Tax Mores," in Donaldson and Werhane, op. cit., pp. 37–39.

Kemper, John Dustin: *The Engineer and His Profession,* 2d ed., Holt, New York, 1975.

Kennan, George F.: "A New Way to Go," Albert Einstein Peace Prize Address, 1981; reprinted in *Fellowship,* Vol. 47, September 1981, pp. 6–8.

Kidder, Tracy: *The Soul of a New Machine,* Avon, New York, 1981.

Kipnis, Kenneth: "Engineers Who Kill: Professional Ethics and the Paramountcy of Public Safety," *Business and Professional Ethics Journal,* Vol. 1, No. 1, 1981, pp. 77–91.

Klein, Heywood, and Hal Lancaster: "Major Flaws Persist in Big Buildings, Often Due to Pressure to Cut Costs," *The Wall Street Journal,* February 12, 1982.

Kleingartner, Archie: "Professionalism and Engineering Unionism," *Industrial Relations,* Vol. 8, May 1969, pp. 224–235.

Kling, Rob: "Computer Abuse and Computer Crime as Organizational Activities," *Computer/Law Journal,* Vol. II, Spring 1980, pp. 403–427.

Kling, Rob: *Social Issues and Impacts of Computing,* University of California, Irvine, 1979.

Kohn, Philip M., and Roy V. Hughson: "Perplexing Problems in Engineering Ethics," *Chemical Engineering,* Vol. 87, No. 9, May 5, 1980, pp. 100–107. Quotations in text used with permission of McGraw-Hill Book Co.

Kotchian, A. Carl: "Lockheed Aircraft Corporation," *Saturday Review,* July 9, 1977; reprinted in Donaldson and Werhane, op. cit., pp. 67–75.

Lachs, John: " 'I Only Work Here': Mediation and Irresponsibility," in Richard T. De George and Joseph A. Pichler (eds.), *Ethics, Free Enterprise, and Public Policy,* Oxford, New York, 1978, pp. 201–213.

Ladd, John: "Loyalty," in Paul Edwards (ed.), *The Encyclopedia of Philosophy,* Vol. 5, Macmillan, New York, 1967, pp. 97–98.

Ladd, John: "The Quest for a Code of Professional Ethics," in Chalk, Frankel, and Chafer, op. cit., pp. 154–159.

Ladenson, Robert F., J. Choromokos, E. d'Anjou, M. Pimsler, and H. Rosen: *A Selected Annotated Bibliography of Professional Ethics and Social Responsibility in Engineering,* Center for the Study of Ethics in the Professions, Illinois Institute of Technology, Chicago, 1980.

Ladenson, Robert F.: "Freedom of Expression in the Corporate Workplace: A Philosophical Inquiry," in J. Ellin, M. Pritchard, and W. Robison (eds.), *Business and Professional Ethics,* Humana Press, Clifton, New Jersey (forthcoming).

Ladenson, Robert F.: "The Social Responsibility of Engineers and Scientists: A Philosophical Approach," in D.L. Babcock and C.A. Smith (eds.), *Values and the*

Public Works Professional, University of Missouri-Rolla, 1980, available from the American Public Works Association, 1313 E. 60 St., Chicago, Illinois 60637.

Lawless, Edward W.: *Technology and Social Shock,* Rutgers, New Brunswick, 1977.

Layton, Edwin T.: "Engineering Ethics and the Public Interest: A Historical View," in Flores, *Ethical Problems in Engineering,* Vol. One, pp. 26–29.

Layton, Edwin T.: *The Revolt of the Engineers,* Case Western Reserve, Cleveland, 1971.

Leiser, Burton M.: "Truth in the Marketplace: Advertisers, Salesmen, and Swindlers," in Burton M. Leiser (ed.), *Liberty, Justice, and Morals,* 2d ed., Macmillan, New York, 1979, pp. 262–297.

Lide, D.R.: "Critical Data for Critical Needs," *Science,* Vol. 212, June 19, 1969, pp. 1343–1349. Figure in text reproduced with permission of author and *Science.* The original appeared in C.Y. Ho, R.W. Powell, and P.E. Liley, *J. Phys. Chem. Ref. Data 3* (Suppl. 1), 1974.

Lockhart, T. W.: "Safety Engineering and the Value of Life," *Technology and Society* (IEEE), Vol. 9, March 1981, pp. 3–5. Quotation in text used with permission of author.

Logsdon, Tom: *Computers and Social Controversy,* Computer Science Press, Potomac, 1980.

Lombardo, Thomas G.: "TMI: An Insider's Viewpoint," *IEEE Spectrum,* Vol. 17, No. 5, May 1980, pp. 52–55.

Lord, Walter: *A Night to Remember,* illustrated edition, Holt, New York, 1976.

Lowrance, William W.: *Of Acceptable Risk,* William Kaufmann, Los Altos, 1976.

Maccoby, Michael: *The Gamesman,* Bantam, New York, 1978.

MacDonald, John D.: *Condominium,* Fawcett, New York, 1977.

Machol, Robert E.: "Principles of Operations Research, 10: The Titanic Coincidence," in *Interfaces* (TIMS/ORSA), Vol. 5, No. 3, May 1975, pp. 53–54.

MacKinnon, Catherine A.: *Sexual Harassment of Working Women,* Yale, New Haven, Connecticut, 1978.

Manchester, William: *The Arms of Krupp, 1587–1968,* Bantam, New York, 1970. Quotations in text used with permission of publisher.

Maner, Walter: "The Management of Information in Political Campaigns," *Computers and Society,* Vol. 11, 1980, pp. 2–9.

Mantell, Murray I.: *Ethics and Professionalism in Engineering,* Macmillan, New York, 1964.

Margolis, Joseph: "Conflict of Interest and Conflicting Interests," in Beauchamp and Bowie, op. cit. pp. 361–372.

Martin, Daniel: *Three Mile Island: Prologue or Epilogue?,* Ballinger, Cambridge, Massachusetts, 1980.

Martin, Mike W.: "Professional Autonomy and Employers' Authority," in Flores, *Ethical Problems in Engineering,* Vol. One, pp. 177–181.

Martin, Mike W.: "Rights and the Meta-Ethics of Professional Morality" and "Professional and Ordinary Morality: A Reply to Freedman," *Ethics,* Vol. 91, July 1981, pp. 619–625 and 631–633.

Martin, Mike W.: "Why Should Engineering Ethics Be Taught?" *Engineering Education,* Vol. 71, No. 4, January 1981, pp. 275–278. Some material from this essay is adapted in Chap. 1 above and used with permission of the American Society of Engineering Education.

Marx, Wesley: *Acts of God, Acts of Man,* Coward, McCann & Geoghegan, New York, 1977. Quotation in text used with permission of author.

Mason, John F.: "The Technical Blow-By-Blow: An Account of the Three Mile Island Accident," *IEEE Spectrum,* Vol. 16, No. 11, November 1979, pp. 33–42.

Mayer, Charles: "Appeals Court Bars Sex Bias by U.S. Firms to Please Foreign Customers," *Los Angeles Times,* Part IV, p. 2, August 20, 1981.

McGregor, Douglas: *The Human Side of Enterprise,* McGraw-Hill, New York, 1960.

McIntyre, Louis V., and Marion Bayard McIntyre: *Scientists and Engineers: The Professionals Who Art Not,* Arcola Communications, Lafayette, 1971.

McQuade, Walter: "Why All Those Buildings Are Collapsing," *Fortune,* November 19, 1979, pp. 58–66.

Melden, A.I.: *Ethical Theories: A Book of Readings,* 2d ed., Prentice-Hall, Englewood Cliffs, New Jersey, 1967.

Melden, A.I.: *Rights and Persons,* University of California, Berkeley, 1977.

Melman, Seymour: "A Note on: Safety Improvement as a Zero Defect Problem," in Flores, *Designing for Safety: Engineering Ethics in Organizational Contexts,* pp. 173–176.

Melman, Seymour: *Pentagon Capitalism,* McGraw-Hill, New York, 1970.

Melville, Herman: *Billy Budd and Other Tales,* New American Library, New York, 1961.

Meyer, Henry Cord: "Politics, Personality, and Technology: Airships in the Manipulations of Dr. Hugo Eckener and Lord Thomson, 1919–1930," *Aerospace Historian,* September 1981, pp. 165–172.

Milgram, Stanley: *Obedience to Authority,* Harper and Row, New York, 1974.

Mill, John Stuart: *Utilitarianism, with Critical Essays,* Samuel Gorovitz (ed.), Bobbs-Merrill, Indianapolis, 1971.

Mironi, Mordechai: "The Confidentiality of Personnel Records," *Labor Law Journal,* Vol. 25, May 1974, pp. 270–292.

Moll, Richard A.: "Product Liability: A Look at the Law," *Engineering Education,* Vol. 66, No. 4, January 1976, pp. 326–331.

Moore, G. E.: *Principia Ethica,* Cambridge, Cambridge, UK, 1966.

Morgan, Arthur E.: *Dams and Other Disasters,* Porter Sargent, Boston, 1971.

Morris, Joe Alex, Jr.: "Computer Age Has Yet to Dawn in Bonn," *Los Angeles Times,* Part I, p. 23, June 15, 1971.

Mostert, Noel: *Supership,* Knopf, New York, 1974.

Murdoch, William W. (ed.): *Environment,* Sinauer Associated, Sunderland, 1975.

Nader, Ralph: "Responsibility and the Professional Society," *Professional Engineer,* Vol. 41, May 1971, pp. 14–17.

Nader, Ralph, Peter J. Petkas, and Kate Blackwell: *Whistle Blowing,* Grossman, New York, 1972.

National Research Council: *Exploratory Study on Responsibility, Liability, and Accountability for Risks in Construction,* National Academy of Sciences, Washington, D.C., 1978.

Nelson, Carl, and Susan Peterson: "Are Cost-Benefit Analyses Immoral?", *ASEE/IEEE Frontiers in Education 1981 Conference Proceedings,* American Society for

Engineering Education, Washington, D.C., and the Institute of Electrical and Electronics Engineers, New York, 1981.

Newhouse, John: *The Sporty Game,* Alfred Knopf, New York, 1982. An earlier version appeared in *The New Yorker* magazines of June 14, 21, 28, and July 5, 1982 as "The Aircraft Industry."

Nixon, F., N.E. Frost, and K.J. March: "Choosing a Factor of Safety," adapted by R.R. Whyte (ed.) for *Engineering Progress Through Trouble* (see below), pp. 136–139. Quotations in text used with permission of the Council of the Institution of Mechanical Engineers.

Nozick, Robert: *Anarchy, State, and Utopia,* Basic Books, New York, 1974.

NSPE Opinions of the Board of Ethical Review, National Society of Professional Engineers, Washington, D.C. Cases are published in the *Professional Engineer* and periodically republished in bound volumes. Quotations in the text used with permission of NSPE.

Oldenquist, Andrew G.: *Moral Philosophy, Text and Readings,* 2d ed., Houghton Mifflin, Boston, 1978.

Oldenquist, Andrew G., and Edward E. Slowter: "Proposed: A Single Code of Ethics for All Engineers," *Professional Engineer,* Vol. 49, May 1979, pp. 8–11.

O'Neill, Brian, and A. B. Kelley: "Costs, Benefits, Effectiveness, and Safety: Setting the Record Straight," *Professional Safety,* August 1975, pp. 28–34.

Otten, James: "Organizational Disobedience," in Flores, *Ethical Problems in Engineering,* Volume One, pp. 182–186.

Papanek, Victor: *Design for the Real World,* Pantheon, New York, 1971.

Parker, Donn B.: *Ethical Conflicts in Computer Science and Technology,* AFIPS Press, Arlington, 1979. Case studies adapted in text with permission of author and publisher.

Perrucci, Robert, and Joel E. Gerstl: *The Engineers and The Social System,* Wiley New York, 1969.

Perrucci, Robert, and Joel E. Gerstl: *Profession Without Community: Engineers in American Society,* Random House, New York, 1969.

Perry, Tekla S.: "Five Ethical Dilemmas," *IEEE Spectrum,* Vol. 18, No. 6, June 1981, pp. 53–60. Quotations in text used with permission of author and the Institute of Electrical and Electronics Engineers.

Peters, Charles, and Taylor Branch: *Blowing the Whistle,* Praeger, New York, 1972.

Pichler, Joseph A.: "Power, Influence and Authority," in Joseph W. McGuire (ed.), *Contemporary Management,* Prentice-Hall, Englewood Cliffs, New Jersey, 1974, pp. 400–434.

Popper, Norman N.: "Trade Secrets: How They Affect Your Job Mobility," *Chemical Engineering,* April 7, 1980, pp. 101–104.

Press, Robert M.: "Southern Florida Alarmed by Drought, Studies Handling of Water Resources," *Los Angeles Times,* Part I-C, pp. 1 and 11, December 1981.

Rabow, Gerald: "The Value of Human Lifetime—and its Application to Environmental and Energy Policy," *Technology and Society* (IEEE), Vol. 9, March 1981, pp. 5–7.

Ramo, Simon: *The Future Role of Engineering,* TRW Inc., Corporate Public Relations, Cleveland, 1976.

Rand, Ayn: *The Virtue of Selfishness,* New American Library, New York, 1964.

Ransom, W. H.: *Building Failures: Diagnosis and Avoidance,* E. & F.N. Spon, London, 1981.

Rasmussen, Norman C.: *Reactor Safety Study,* U.S. Atomic Energy Commission, WASH 1400, Draft of August 1974 (known as the "Rasmussen Report").

Raven-Hansen, Peter: "Dos and Don'ts for Whistleblowers: Planning for Trouble," *Technology Review,* Vol. 82, May 1980, pp. 34–44.

Rawls, John: *A Theory of Justice,* Harvard, Cambridge, Massachusetts, 1971.

Reed, George L.; "Moonlighting and Professional Responsibility," *Journal of Professional Activities, Proceedings of the American Society of Civil Engineers,* Vol. 96, September 1970, pp. 19–23.

Regan, Tom: *All That Dwell Therein: Essays on Animal Rights and Environmental Ethics,* University of California Press, Berkeley, 1982.

Reiman, Jeffrey H.: "Privacy, Intimacy, and Personhood," in Richard A. Wasserstrom (ed.), *Today's Moral Problems,* Macmillan, New York, 1979, pp. 377–391.

Rescher, Nicholas: *Unpopular Essays on Technological Progress,* University of Pittsburgh, Pittsburgh, 1980.

Riegel, J.W.: *Collective Bargaining as Viewed by Unorganized Engineers and Scientists,* University of Michigan, Ann Arbor, 1959.

Rivlin, Alice M.: *Systematic Thinking for Social Action,* The Brookings Institution, Washington, D.C., 1971.

Robinson, Douglas H.: *Giants in the Sky,* University of Washington Press, Seattle, 1973.

Roche, James M.: "The Competitive System, To Work, To Preserve, and To Protect," *Vital Speeches of the Day,* May 1, 1971, p. 445.

Rogovin, Mitchell, and George T. Frampton, Jr.: *Three Mile Island, A Report to the Commissioners and The Public,* Vol. I, Nuclear Regulatory Commission Special Inquiry Group, NUREG/CR-1250, Washington, January 1980. Diagram in text (Fig. 4-12) used with permission of Mitchell Rogovin.

Rosenbaum, Walter A.: *The Politics of Environmental Concern,* 2d ed., Praeger, New York, 1977.

Ross, W.D.: *The Right and the Good,* Oxford, Oxford, 1946.

Ross, Steven S.: "Technical Illiteracy," *New Engineer,* March 1978, p. 6. Quotation in text used with permission of author-editor.

Rowe, William D.: *An Anatomy of Risk,* Wiley, New York, 1977.

Rowe, William D.: "What is an Acceptable Risk and How Can It Be Determined?" in G.T. Goodman and W.D. Rowe (eds.), *Energy Risk Management,* Academic, 1979, pp. 327–344.

Rule, James, Douglas McAdam, Linda Stearns, and David Uglow: *The Politics of Privacy,* New American Library, New York, 1980.

Sagan, L.A.: "Human Cost of Nuclear Power," *Science,* Vol. 177, August 11, 1972, pp. 487–493.

Samuelson, Robert J.: "Industrial Espionage—Not to Worry," *Los Angeles Times,* Part II, p. 13, July 15, 1982.

Sayre, Kenneth (ed.): *Values in the Electric Power Industry,* University of Notre Dame, Notre Dame, 1977.

Schinzinger, Roland: "The Engineer as an Agent of Change" (unpublished manuscript, 1973).

Schinzinger, Roland, and Mike W. Martin: "Engineering as Social Experimentation," in *1980 ASEE Annual Conference Proceedings,* Vol. 2, American Society
for Engineering Education, Washington, D.C., 1980, pp. 394–398.

Schinzinger, Roland: "The Experimental Nature of Engineering and Its Implications
for Management," *Technology and Society* (IEEE), Vol. 7, No. 27, September
1979, pp. 3–5.

Schinzinger, Roland, and Mike W. Martin: "The Experimental Nature of Engineering
and Its Implications for the Style of Engineering Practice," in *1980 Frontiers in
Education Conference Proceedings (Houston),* American Society for Engineering
Education, Washington, D.C., and The Institute of Electrical and Electronics
Engineers, New York, 1980, pp. 204–207.

Schwartz, Eugene S.: *Overskill,* Quadrangle, Chicago, 1971.

Schwarze, Sharon: "Intellectual Property and the Justification of Intellectual Property Rights" (unpublished manuscript).

Seidman, Joel: "Engineering Unionism," in Robert Perrucci and Joel E. Gerstl (eds.),
The Engineers and The Social System, Wiley, New York, 1969, pp. 219–245.

Seldes, George: *Iron, Blood, and Profits,* Harper and Brothers, New York, 1934.

Sethi, S. Prakash: *Up Against the Corporate Wall,* 3d ed., Prentice-Hall, Englewood
Cliffs, New Jersey, 1977.

Shapley, Deborah: "Unionization: Scientists, Engineers Mull over One Alternative,"
Science, Vol. 176, May 12, 1972, pp. 618–621.

Shapo, Marshall S.: *A Nation of Guinea Pigs,* Free Press, New York, 1979. Quotation
in text used with permission of Macmillan Publishing Company.

Shaw, Gaylord: "Bureau of Reclamation Harshly Criticized in New Report on Teton
Dam Collapse," *Los Angeles Times,* June 4, 1977, Part I, p. 3.

Shedd, John A.: *Salt from My Attic,* The Mosher Press, Portland, Me., 1928, p. 20.

Shue, Henry: "Exporting Hazards," *Ethics,* Vol. 91, July 1981, pp. 579–606.

Shute, Nevil: *No Highway,* Charter, New York, 1976.

Silverman, Milton, P. Lee, and M. Lydecker: *The Drugging of the Third World,*
Institute for Health Policy Studies, University of California, San Francisco,
1981.

Simon, Herbert A.: *Administrative Behavior,* 3d ed., Free Press, New York, 1976.

Simrall, Harry C.: "The Civic Responsibility of the Professional Engineer," *The
American Engineer,* May 1963, pp. 39–40.

Slovic, Paul, Baruch Fischhoff, and Sarah Lichtenstein: "Rating the Risks,"
Environment, Vol. 21, April 1979, pp. 14–20 and 36–39.

Slovic, Paul, Baruch Fischhoff, and Sarah Lichtenstein: "Risky Assumptions,"
Psychology Today, Vol. 14, No. 1, June 1980, pp. 44–48. Quotations in text used
with permission of Ziff Davis Publishing Co.

Slovic, Paul, Baruch Fischhoff, and Sarah Lichtenstein: "Weighing the Risks: Which
Risks are Acceptable?" *Environment,* Vol. 21, May 1979, pp. 17–20 and 32–38.

Smith, Adam: *The Wealth of Nations,* University of Chicago, Chicago, 1976.

Smith, Peter: "Designer Interest," in *Exploratory Study on Responsibility, Liability,
and Accountability for Risks in Construction,* National Research Council, National Academy of Sciences, Washington, D.C., 1978, pp. 31–35.

Smith, R. Jeffrey: "Court Upholds Controversial Regulations," *Science,* Vol. 213,
July 10, 1981, pp. 185–188. Quotation in text used with permission of author.

Soderberg, C. Richard: "The American Engineer," in Kenneth S. Lynn (ed.), *The Professions in America*, Beacon, Boston, 1967, pp. 203–230.

Sowers, George B., and George F. Sowers: *Introductory Soil Mechanics and Foundations*, 3d ed., Macmillan, New York, 1970.

Starr, Chauncey, Richard Rundman, and C. Whipple: "Philosophical Basis for Risk Analysis," *Annual Review of Energy*, Vol. 1, 1976, pp. 629–662. Graph in text used with permission of the *Annual Review of Energy*.

Starr, Chauncey: "Social Benefit Versus Technological Risk," *Science*, Vol. 165, September 19, 1969, pp. 1232–1238. Graph in text used with permission of author.

Stephens, Mark: *Three Mile Island*, Random House, New York, 1980.

Stevenson, Charles L.: "Persuasive Definitions," *Mind*, Vol. 47, 1938, pp. 331–350.

Stockholm International Peace Research Institute: *World Armaments and Disarmament*, SIPRI Yearbook 1980, Taylor & Francis, London, 1980; Crane, Russak & Co., New York, 1980.

Stone, Christopher D.: *Where the Law Ends: The Social Control of Corporate Behavior*, Harper and Row, New York, 1975.

Storch, Laurence: "Attracting Young Engineers to the Professional Society," *Professional Engineer*, Vol. 41, May 1971, pp. 36–39.

Sugarman, Robert: "Nuclear Power and the Public Risk," *IEEE Spectrum*, Vol. 16, No. 11, November 1979, pp. 59–79.

Swenson, Gerald S.: "The Case for Fee Competition," *Consulting Engineer*, Vol. 50, June 1978, pp. 90–96.

Taylor, Paul W.: *Principles of Ethics, An Introduction*, Dickenson, Encino, California, 1975.

Thorpe, James F., and William H. Middendorf: *What Every Engineer Should Know About Product Liability*, Marcel Dekker, New York, 1979.

Thrall, Charles A., and Jerold M. Starr (eds.): *Technology Power, and Social Change*, Lexington, Lexington, Massachusetts, 1972.

Tversky, Amos, and Daniel Kahneman: "The Framing of Decisions and the Psychology of Choice," *Science*, Vol. 211, January 30, 1981, pp. 453–458.

Ui, Jun (ed.): *Polluted Japan*, Jishu-Koza Citizens' Movement, Tokyo, 1972.

Unger, Stephen H.: *Controlling Technology: Ethics and The Responsible Engineer*, New York, Holt, Rinehart and Winston, 1982.

Unger, Stephen H.: "How to be Ethical and Survive," *IEEE Spectrum*, Vol. 16, No. 11, December 1979, pp. 56–7.

Vandivier, K.: "Engineers, Ethics and Economics," in *Conference on Engineering Ethics*, American Society of Civil Engineers, New York, 1975, pp. 20–24.

Vaughn, Richard C.: *Legal Aspects of Engineering*, 3d ed., Kendall/Hunt, Dubuque, 1977. Quotations in text used with permission of publisher.

Veblen, Thorstein: *The Engineers and the Price System*, Viking, New York, 1965.

Velasquez, Manuel G.: *Business Ethics: Concepts and Cases*, Prentice-Hall, Englewood Cliffs, New Jersey, 1982.

Vetter, Betty M.: "Engineering: Science Outlook for Women," *Professional Engineer*, Vol. 50, June 1980, pp. 29–31.

Vonnegut, Kurt, Jr.: *Player Piano*, Dell, New York, 1952.

Wade, Wynn C.: *The Titanic: End of a Dream*, Penguin, New York, 1980.

Walters, Kenneth: "Professionalism and Engineer/Management Relations," *Professional Engineer,* Vol. 43, January 1973, pp. 41–42.

Walters, Kenneth: "Your Employees' Right to Blow the Whistle," *Harvard Business Review,* Vol. 53, July 1975, pp. 26–34.

Walton, Richard E.: *The Impact of The Professional Engineering Union,* Harvard University Press, Cambridge, Massachusetts, 1961.

Weber, Max: *The Protestant Ethic and The Spirit of Capitalism,* Charles Scribner's Sons, New York, 1958.

Weber, Max: *The Theory of Social and Economic Organization,* Talcott Parsons (ed.), Free Press, New York, 1947.

Weil, Vivian: *Action and Responsibility in the Engineering Profession,* C.S.E.P. Occasional Papers No. 2, Center for the Study of Ethics in the Professions, Illinois Institute of Technology, Chicago, 1979.

Weinberg, Alvin M.: "The Many Dimensions of Scientific Responsibility," *Bulletin of the Atomic Scientist,* November 1976, pp. 21–25.

Weinstein, Deena: *Bureaucratic Opposition,* Pergamon, New York, 1979.

Westin, Alan F., and Stephan Salisbury (eds.): *Individual Rights in the Corporation,* Random House, New York, 1980.

Westin, Alan F.: *Privacy and Freedom,* Atheneum, New York, 1967.

Westin, Alan F. (ed.): *Whistle-Blowing!: Loyalty and Dissent in the Corporation,* McGraw-Hill, New York, 1981.

Whitelaw, Robert L.: "The Professional Status of the American Engineer: A Bill of Rights," *Professional Engineer,* Vol. 45, August 1975, pp. 37–41.

Whiteside, Thomas: *Computer Capers,* Crowell, New York, 1978.

Whyte, R.R. (ed.): *Engineering Progress Through Trouble,* The Institution of Mechanical Engineers, London, 1975. Quotations in text used with permission of the Council of the Institution of Mechanical Engineers.

Whyte, William H.: *The Organization Man,* Simon and Schuster, New York, 1956.

Wiener, Norbert: "A Scientist Rebels," *Atlantic Monthly,* Vol. 179, January 1947, p. 46.

Wiener, Norbert: *God and Golem, Inc.,* MIT, Cambridge, Massachusetts, 1964.

Wiesner, Jerome B., and Herbert F. York: "The Test Ban," *Scientific American,* Vol. 211, 1964, pp. 27–35.

Wildavsky, Aaron: "No Risk Is the Highest Risk of All," in Albert Flores (ed.), *Ethical Problems in Engineering,* pp. 221–226.

Wohl, Burton: *The China Syndrome,* Bantam, New York, 1979.

Wright, J. Patrick: *On a Clear Day You Can See General Motors,* Avon, New York, 1979.

"Yarrow Bridge," Editorial, *The Engineer,* Vol. 210, October 23, 1970, p. 415.

NAME INDEX

SUBJECT INDEX